An Introduction to
Soil Reinforcement and Geosynthetics

G L Sivakumar Babu

Universities Press

Universities Press (India) Private Limited

Registered Office
3-6-747/1/A & 3-6-754/1, Himayatnagar,
Hyderabad 500 029 (A.P.), India
e-mail: info@universitiespress.com

Distributed by
Orient Blackswan Private Limited

Registered Office
3-6-752, Himayatnagar, Hyderabad 500 029 (A.P.), India

Other Offices
Bangalore / Bhopal / Bhubaneshwar / Chennai / Ernakulam
Guwahati / Hyderabad / Jaipur / Kolkata / Lucknow / Mumbai
New Delhi / Patna

First published 2006
Reprinted 2009

ISBN: 978 81 7371 481 8

Set in Times 10 on 12 by
AICRA Info Mates, Hyderabad 500 082

Printed in India at
Orion Printers Private Limited
Hyderabad 500 004

Published by
Universities Press (India) Private Limited
3-6-747/1/A & 3-6-754/1, Himayatnagar
Hyderabad 500 029 (A.P), India

Contents

Preface

The modern concept of soil reinforcement and the use of geosynthetics in civil engineering evolved about thirty years ago. Since then, these materials have been put to a spectacular range of applications thanks to the simplicity of the design principles involving them and the ease of their applicability to construction practices. Geosynthetics have also gained widespread acceptance among civil engineering professionals as an economic and innovative alternative to conventional materials to such an extent that, nowadays, in some developed and developing countries, engineering designs implicitly assume the use of these materials in one form or the other.

Credit for this goes to Henry Vidal, Robert Koerner and a number of other distinguished practitioners of this technology in civil engineering. Robert Koerner rightly asserts that geosynthetics have a place comparable to soil, concrete and steel in civil engineering industry. The developments in this field have been so rapid and multifaceted that to present a comprehensive coverage of them is a difficult task. In this book I have attempted to provide an introduction to the topic. Our country is on the threshold of massive expansion on the infrastructure front; many national and multinational companies are and will be involved in the designing, construction and execution of building projects. A primer on the role of geosynthetics and soil reinforcement in this construction endeavour will be timely, and I hope this book will fulfill the requirement. The works of R. M. Koerner, J. P. Giroud and R. A. Jewell in the area have been inspirational, and the material presented in this volume is based on the contributions of these distinguished practitioners.

The first chapter deals with the initial and recent developments in the field of soil reinforcement and the techniques pertaining to the same. The second chapter deals with the design principles, influencing factors, mechanisms, soil–reinforcement interactions, typical specifications for backfill soils, etc. The third chapter looks at different types of reinforcement materials such as steel and geosynthetics, and evaluation of their material properties. The improvement of bearing capacity of footings on sands and clays using soil reinforcement technique, reinforcement of slopes to increase stability, and the design of reinforced soil retaining walls are discussed in the fourth, fifth and sixth chapters, respectively. The seventh chapter deals with construction of embankments on soft soils using reinforcement techniques. Soil nailing is an established technique for improving stability of cuts and slopes. The advantages of this technique, various design methods and practices of soil nailing are discussed in detail in the eighth chapter. Chapter nine covers the use of geosynthetics in filtration and drainage, while the use of geosynthetics in roads (unpaved and paved) is discussed in Chapter 10. Chapter 11 covers the use of geosynthetics in landfills; specifications of various kinds of landfills, their applications, advantages and concerns regarding their use are also given here. Chapter twelve looks at the use of natural materials such as jute, coir and bamboo for filtration, separation, drainage and reinforcement functions. In all the chapters, examples have been provided wherever possible to illustrate the principles; they do not directly form design solutions.

Soil reinforcement and geosynthetics is a dynamic subject and its influence extends to professional activities in engineering, teaching and research alike. The use of metallic and geosynthetic materials is on the rise and so are the new-found areas of their application. The scope for innovation here is unlimited and therein lies the excitement and the challenge.

Acknowledgements

I gratefully acknowledge the financial support provided by the Centre for Continuing Education (CCE), IISc, Bangalore, in the preparation of course material. Considerable part of the material presented here was collated during my stay (1999–2000) in Germany as an Alexander von Humboldt Fellow; I thank the Alexander von Humboldt Foundation, Germany, Dr Erwin Gartung and Mr. Helmut Zanzinger of the Geotechnical Institute, Nuremberg, for all their help.

I am thankful to my teachers and colleagues Prof. N. S. Pandian, Prof. B. R. Srinivasa Murthy, Prof. T. S. Nagaraj, Prof. A. Sridharan and Prof. K. S. Subba Rao for their encouragement and guidance at various stages of my career. I have enjoyed interacting with Prof. T. G. Sitharam, Prof. M. M. Allam, Prof. M. Sudhakar Rao, Prof. P. V. Sivapullaiah, Dr. Jyant Kumar, and Mr P. V. Raghuveer Rao and thank them for their useful comments. I also thank Dr. R. Seshagiri Rao, Mr. D. Sathyanarayana Murthy, and Mr. M. D. Mukesh for their help in the preparation of the manuscript. I thank my wife K. Himabindu and children G. Deepti and G. Vennela for their love and support.

G. L. Sivakumar Babu
Bangalore

1 Introduction

1.1 Initial Developments

From the time immemorial people have been aware that tree roots strengthen soil. People in ancient times used bamboo, wood and other materials along with sand and mixtures of mud and clay in house construction. The use of logs or tree trunks of uniform sizes, fixed together to form a mattress for crossing marshy areas, is quite old (Koerner, 1998). In India, it has been customary to add a little amount of fibre while making idols of Lord Ganesha and other deities for increasing the workability of the material and reducing shrinkage cracks. Jones (1996) makes a reference to the use of reinforcement in the constructions of the city of Baghadad, the seat of Mesopotamian civilization, and Great Wall of China as the earliest known. He also indicates that structures such as reed-reinforced earth levees, reinforced wharves using timber elements embedded in backfill, fortifications using alternate layers of wooden logs and earth fills, river training works through dykes, etc., reported in the construction works of older days involved soil reinforcement principle. An important contribution was made to the use of reinforced soil in military works in 1882, when Colonel Pasley of the British army made use of alternate horizontal layers of brushwood, wooden planks and canvas in the backfill. He deduced that such an arrangement, in combination with soil, reduces lateral pressures significantly. Reinforcement techniques were employed to strengthen the downstream face of dams in 1904 in California; similar attempts were made in constructing retaining walls in USA in 1925 and in France in 1927.

1.1.1 Geosynthetics

'Geosynthetics' is a term of recent origin which essentially includes the family of materials that are used in conjunction with soil to improve its performance in a specific context. It includes geotextiles, geomembranes, and a host of other products such as geogrids, geonets, geocomposites, geosynthetic clay liners, geopipes, geobags, etc. They have been used all over the world in a number of applications such as construction of road separators; reinforcement of embankments, slopes and retaining walls; moisture/vapour barriers; and in erosion control and filtration.

According to John (1987), the earliest reported use of such material is in the form of woven cotton fabrics in 1926 by the South Carolina Highways Department. A fabric covered with hot asphalt was used during road construction; its function was similar to a geomembrane. Some of the other earliest known applications of geosynthetics include their use beneath concrete block revetments in 1957 and again in 1958 in the waterfront structures in the state of Florida, USA. They were also used in Holland in 1960 for coastal protection work in place of granular filters. The development of geotextiles received a big stimulus when millions of square metres of the material was used by the Dutch Delta Works

Scheme for improving the coastal defenses along the rivers Rhine, Maas and Schelde following severe storms in Holland (John, 1987).

1.2 Recent Developments

1.2.1 Reinforced soil

Henry Vidal (Fig. 1.1*a*) introduced the modern concept of reinforced soil in the 1960s. The idea came from a heap of sand reinforced with pine needles (Fig. 1.1*b*) which his daughter had made while playing on the beach. Figure 1.2*a* shows the first retaining wall built in France in 1968 using Vidal's concept, employing U-shaped metal channel members for facing. Following this there was then a spurt of construction activities using reinforced earth technology in many countries such as France, United Kingdom, Japan and USA. Figure 1.2*b* shows the type of construction being followed today using cruciform construction. To date, over 7,500 projects and 65 million square feet of reinforced earth walls have been built in the United States by the Reinforced Earth Company alone using this technique, providing cost savings of over 800 million dollars. More than 300,000 such structures have been build worldwide. If one counts the use of geosynthetics and other forms of reinforcement such as steel bars, the number is likely to be three to four times more. A few innovations in the form of multi-anchor system using anchor plates were also made during the 1980s. Similar attempts using Z-type anchors and loop systems followed in UK, Japan and USA. Fukuoka (1980) and Brandl and Dalmatier (1986) developed anchor systems shown in Fig. 1.3*a* and Fig. 1.3*b*, which have been used in large-scale applications in Japan and Europe. The advantage with anchor systems is that the development of additional passive resistance in the soil reduces the total length of reinforcement required as compared to earlier design concepts. Soil nailing, which was developed in Germany and Europe essentially to

(a) (b)

Fig. 1.1 *(a) Henry Vidal; (b) reinforced sand heap.*

facilitate the driving of tunnels, became a very useful technique in the construction of excavations and cut slopes. The above developments were very significant and led to the use of reinforced soil technique in many applications such as increasing bearing capacity, stabilizing slopes, building retaining walls and embankments on soft soil, pavements, etc. Alternative and economical designs are possible for every conceivable application of reinforcement, with imagination being only the limiting factor in the range of applications.

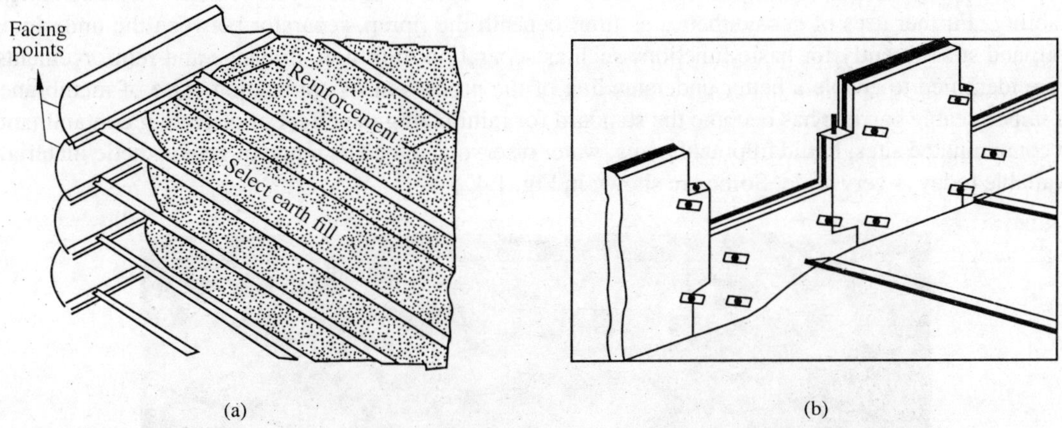

Fig. 1.2 *(a) First retaining wall built in France in 1968 using Vidal's concept using U-shaped metal channel members; (b) modern retaining wall using cruciform construction.*

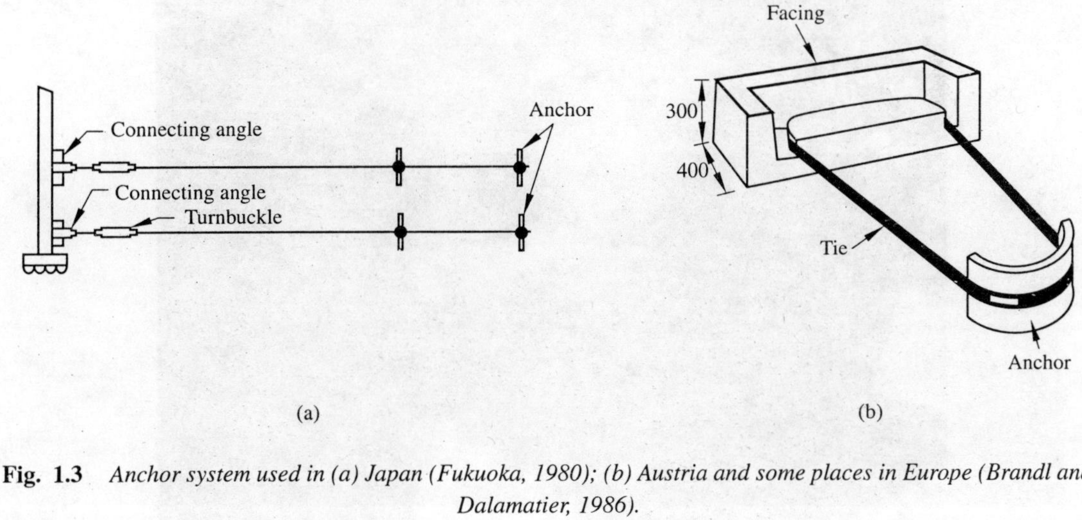

Fig. 1.3 *Anchor system used in (a) Japan (Fukuoka, 1980); (b) Austria and some places in Europe (Brandl and Dalamatier, 1986).*

1.2.2 Geosynthetics

Systematic application of geotextiles did not immediately follow the initial developments that took place from 1930s to 1960s; only when parallel developments took place in materials technology increasing its availability, capabilities and cost effectiveness, did its use become popular. The polyethylene nets manufactured by Netlon Ltd. UK, were used in Japan for reducing the damage to embankment caused by seismic activity and rainfall. They were used as basal layers to improve embankment stability. Further uses of geosynthetics as filter beneath the riprap, separator between the underlying soil, and subsequently for basic functions such as separation, filtration, drainage and reinforcements, were identified to enable a better understanding of the process of its application. Use of membranes as impermeable barriers has become the standard for minimizing the effect of migrating contaminants in contaminated sites, liquid impoundments, water reservoirs, etc. The range of geosynthetic materials available today is very wide. Some are shown in Fig. 1.4.

Fig. 1.4 *Varieties of geosynthetic materials in use (Koerner, 1998)*

Giroud (1986) brings out several reasons as to why these materials have become successful and acceptable in professional practice. Three groups of people, viz., contractors, designers and owners are behind this spectacular growth. Contractors are interested in quicker, less weather-dependent construction materials, reduced volume of earthworks and capability of handling poor quality soil as well. Designers want better reliability, flexibility and control over their designs, which are possible with the use of appropriate choice of geotextile materials. Owners expect cost savings in both construction and maintenance. Giroud suggests the following reasons for the wide acceptability of geosynthetics.

1. The membrane-like behaviour of these materials complement the properties of granular soil. Hence, soils that are prone to erosion or settlement can be stabilized. The tensile properties of geosynthetic material become useful in increasing the tensile strength of soils. In addition, these materials are effective as a separator between dissimilar layers of soil.
2. Geotechnical structures such as pavements, foundations and retaining walls use layered construction. Geosynthetic materials are used as interfaces between layers. This improves the overall performances.
3. Geotechnical structures need to be flexible when subjected to differential settlements. Geosynthetic materials, owing to their inherent flexibility, are compatible with the behaviour of geotechnical structures.

1.3 Applications and Potential

The use of geosynthetics in soil reinforcement and other applications is widely practised in western countries. If the potential of these materials is rightly understood, the use of techniques employing geosynthetics is bound to increase significantly in many developing countries. While the reinforcement function is possible using steel/polymeric reinforcement in the form of grids, rods and strips, functions such as separation, filtration, drainage and moisture barrier are possible only using geosynthetics.

1.3.1 Soil reinforcement function

The soil reinforcement function ensures improved response of the soil to loading. It increases normal loading and reduces disturbing shear forces. As a result steep slopes can be constructed or an embankment of increased height be built on soft soil, which otherwise would not have been possible. In pavements, geosynthetics, owing to their membrane action which induces tensile forces, control deformations and significantly improve their performance. They have similar applications in areas having sinkhole formations or in pile caps in a piled embankment in soft soil. Figures 1.5(*a–c*) shows some applications. Figure 1.5*d* shows the use of reinforcement to stiffen the foundation beneath a reinforced soil retaining wall. Figure 1.6 shows some applications of geotextiles: as soil reinforcment in vertical walls and steep slopes; in slip prevention and repair; and in embankment on soft soil. Some of these aspects are discussed more elaborately in the subsequent sections, and their mechanisms and designs for specific applications are provided.

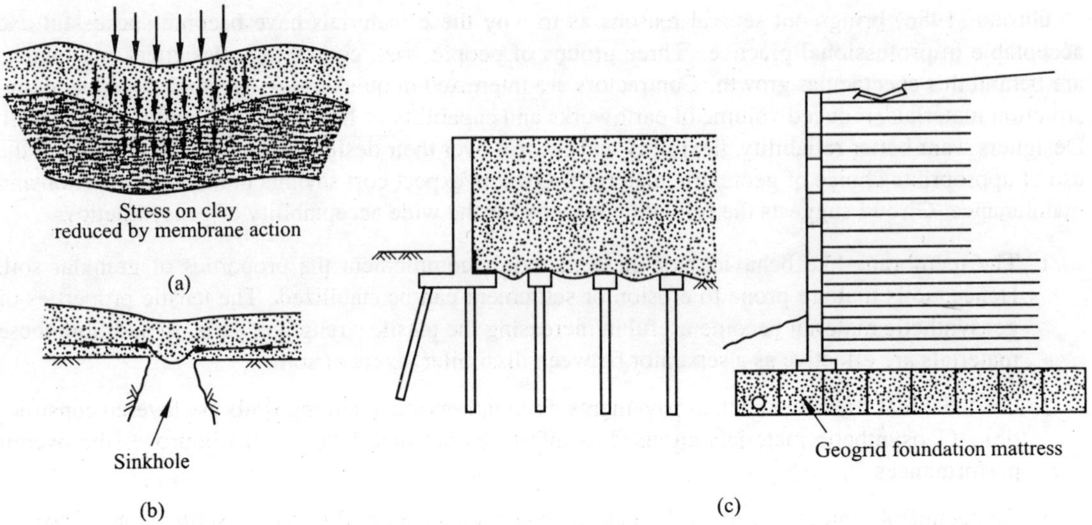

Fig. 1.5 *Reinforcement supporting load by membrane action: (a) rutted unpaved road, (b) above a sink hole, and (c) piled embankment foundation, (d) use of reinforcement to stiffen the foundation beneath a reinforced soil.*

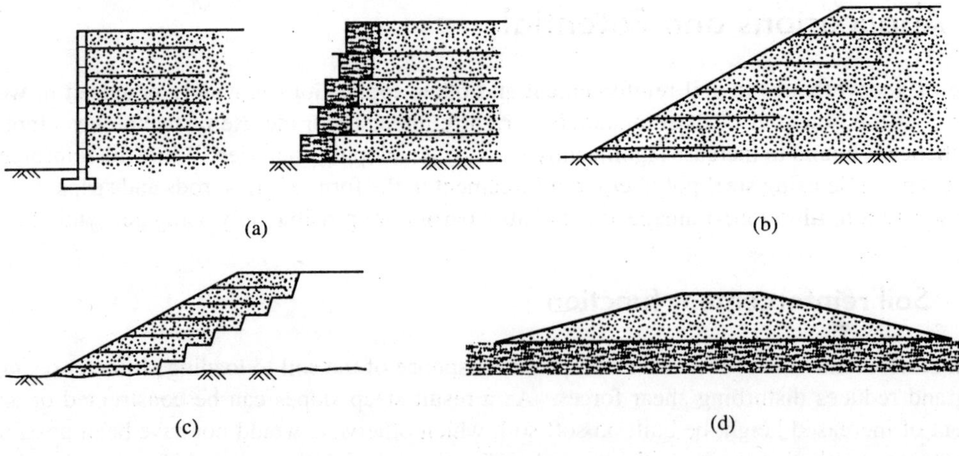

Fig. 1.6 *Example applications of geotextile as soil reinforcement: (a) vertical walls, (b) steep slopes, (c) slip prevention and repair, and (d) embankment on soft soil (Jewel, 1996).*

1.3.2 Separator function

Geosynthetics is useful to separate dissimilar soils (Fig. 1.7). A separator helps prevent intermixing of poor in situ soil with good quality granular materials when the structure is subjected to surface loading. It acts as a barrier to migration of particles between two dissimilar soils, a requirement in many applications such as pavements, graded filters, clay covers and liners in landfills, etc. The separator needs to have sufficient strength to survive ripping, puncturing on impact, and must be durable against chemical attacks.

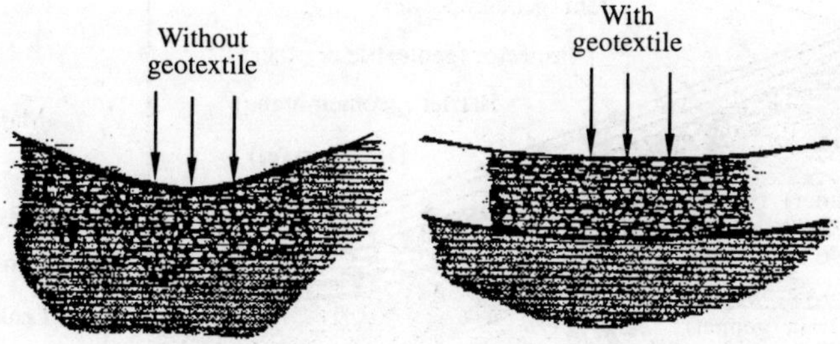

Fig. 1.7 *Geotextile separation function*

1.3.3 Filtration function

Geosynthetics such as geotextiles allow the movement of liquid across the plane of the geotextile while retaining the soil on the upstream side. To achieve this, the material should meet condition of both adequate permeability, requiring an open fabric structure, and soil retention requiring a small opening size. The long term soil-to-geotextile flow compatibility is also important in that the filter must not clog excessively during its lifetime. The filtration function of geotextile is a major area of importance in many applications such as retaining walls, pavements, and in erosion control measures and landfills.

1.3.4 Drainage function

The drainage function of geotextiles enables flow in the plane of the geotextile. Apart from geotextiles that transmit fluids, nowadays, there are specific products such as geonets and drainage composites to meet drainage requirements. Other than the ability to direct the flow, geotextiles should be able to retain soil. Geosynthetic drains have been successful in many cases since they reduce the consolidation time in soft soils. Nowadays, drainage composites are available which serve dual functions of filtration and drainage; some of these products are used extensively in landfills (Fig. 1.8).

1.3.5 Barrier function

The barrier function using geomembranes and geotextiles coated with bitumen, rubber bitumen or other polymer mixes is useful in many applications for minimising fluid and vapour flow. The fluid permeability of materials is very low and it is therefore useful in many applications such as canal linings, liquid impoundments, dams, municipal waste landfills, among others. The materials, to some extent, also prevent the flow of gases and odours that are likely to emanate from solid waste landfills. Figure 1.8 shows the uses of geosynthetics as separator, barrier (geomembrane and geosynthetic clay liner) and drainage layer (geocomposite).

In many of the applications, though the improvement in behaviour or performance is possible primarily through one of the mechanisms discussed above, it is likely that other mechanisms may also

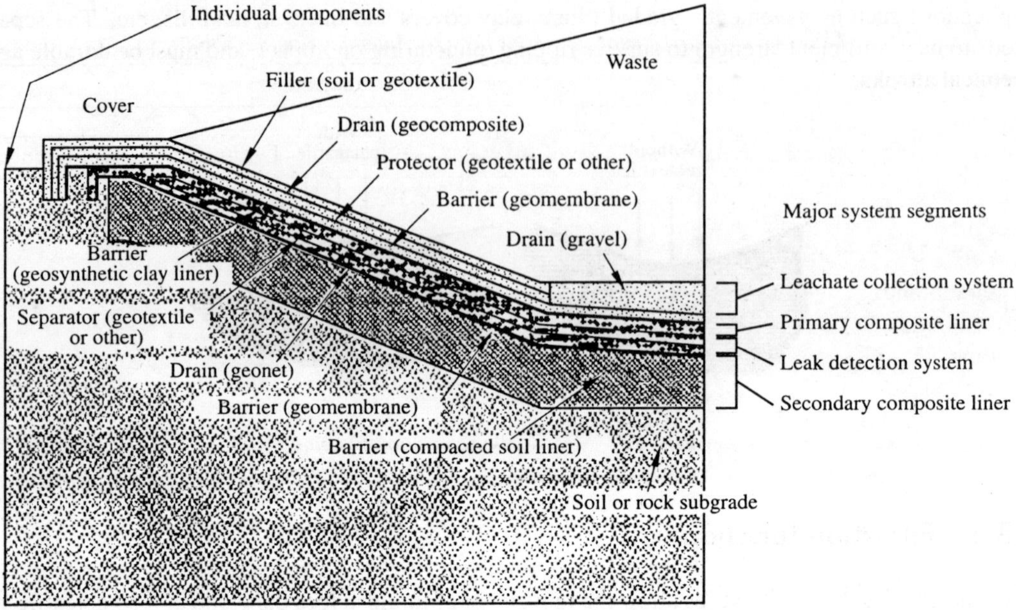

Fig. 1.8 *Uses of geosynthetics in the form of separator, filter, drainage layer, protector and barrier in an engineered landfill*

play an equally important role. For example, in pavements, separation, reinforcement, filtration and drainage are involved. In asphalt overlays, geomembranes are used both for its moisture barrier and reinforcement functions. Table 1.1 illustrates the role and importance of basic functions of geosynthetics. Koerner (2000) presents seventeen different applications involving geosynthetics that have been categorized as geotechnical, transportation, hydraulics and geoenvironmental. The application of each category is mentioned along with its established use as well as future possibilities in Table 1.2.

1.4 Advantages of the Techniques

It is necessary to put into proper perspective the savings in costs achieved and the additional benefits that the use of these materials have resulted in, resulting in its popularity. The benefits have been categorized as economic and technical (Jewell, 1996).

(a) Economic benefits results from:

- Reduced earthwork quantities as a result of steeper slopes being possible and less land acquisition costs.
- Increased construction speed as the reinforcements and facings are made available at the site.
- Unrestricted access that is made possible by the reinforced soil being stable throughout the construction; it occupies less space.
- The usability of poor fill materials with strengthening.

Table 1.1 *Geosynthetics vis-à-vis primary function (Koerner, 2000)*

Types of geosynthetic material	Primary function				
	Separation	Reinforcement	Filtration	Drainage	Containment
Geotextiles	Yes	Yes	Yes	Yes	No[a]
Geogrids	No[b]	Yes	No	No	No
Geonets	No	No	No	Yes	No
Geomembranes	Yes[c]	No[d]	No	No	Yes
Geosynthetic clay liners	Yes[c]	No[d]	No	No	Yes
Geopipe	No	No	No	Yes	No
Geocomposites	a/d[e]	a/d[e]	a/d[e]	a/d[e]	a/d[e]

[a] Unless impregnated by bitumen or other polymer [b] Unless very large particle size soils are involved [c] Usually considered to be secondary function [d] Unless fabric reinforced and designed accordingly [e] Application-dependent design of particular product

(b) Technical benefits arise from:

- Efficient or synergetic use of material combinations, such as shearing resistance of the soil with tensile capacity of the reinforcement.
- Inherent flexibility and tolerance to deformation of a reinforced soil structure. This observation is particularly important in areas of seismic activity. For example, in Japan, a country of considerable seismic activity, it was observed that the retaining walls made using reinforced soil construction withstood severe earthquakes, whereas those constructed with conventional construction using reinforced concrete collapsed.
- Critical components such as facing, reinforcement and connection as a result of strict quality control, and backfill soil due to good compaction control, are likely to contribute significantly to the overall improvement of performance.
- In many cases the techniques offer totally different and new solutions, which are otherwise not possible.

Table 1.2 *Applications along with emerging developments and future possibilities (Koerner, 2000)*

Category	Topic description	Emerging development	Future possibilities
Geotechnical	Soil slope reinforcement	Composite geosynthetics to provide internal drainage	Electrokinetic geosynthetics to modify backfill soils in situ
	Walls and bulkheads	New facing systems, particularly those with vegetative growth, i.e., "live walls"	Polymer rope, straps, or anchors for unique site situations

Contd.

Table 1.2 *Contd.*

Category	Topic description	Emerging development	Future possibilities
	Base reinforcement over voids	Resist differential settlement	Prestress the reinforcement to eliminate all deformation
	Concrete dam waterproofing	Underwater installation and bubbling systems for ice breakers	Improved polymer formulations to resist harsh exposed conditions
	Earth dam waterproofing	Extension into roller-compacted concrete dams	Remediation of existing dam embankment and foundations
	Tunnel waterproofing	Simultaneous installation along with permanent liner system	Improved polymer formulations to provide extremely long service life
Transportation	Geosynthetic modified roadways	Fibre reinforcement and microgrids	Pothole repair using geosynthetic drainage materials
	Trenchless pipe remediation	Pipe bursting to increase diameter of host pipe	Remote entry and robotic controlled connections
	Erosion and control systems	Reinforced geosynthetics and soil nailing for slope stabilization	Avalanche control nets
Hydraulics	Canal liners	Lining of canals under flowing water	Improved polymer formulation to resist exposed conditions
	Geotextile tubes for erosion control	Rugged (high strength) fabrics to resist vandalism and ultraviolet resistance	Low friction, pervious, inner liners to accommodate long flow lengths
	Geotextile containers for soil disposal	Stacked containers for development of underwater structures	Formation of underwater habitats
	Agriculture liners	Development of protective surface for ruggedness during maintenance	Diffusion released cleaning agents within the polymer formulation
Geoenviron-mental	Landfill liner systems	Leachate recycling	Landfill mining
	Landfill closure systems	Gas capture and utilization for power generation	Socially responsive and worthwhile final uses
	Vertical barriers for abandoned dumps	Creation of a gate for focused release of leachate for subsequent treatment	Utilizing double barrier systems with leak detection
	Bottom barriers for abandoned dumps	Directional drilling or jet grouting	Longwall mining installation methods

Though long term performance, durability and non-availability of codes in certain areas are issues of some concern regarding geosynthetics, the use of these materials has become indispensable in civil engineering practice. With regard to durability and long-term performance, it can be stated that the behaviour of steel is known to civil engineering profession, and adequate sacrificial thickness around the reinforcement takes care of its long-term performance in adverse soil conditions. Design methods for geosynthetic materials using adequate partial factors of safety considering time, temperature, creep effects and installation effects are well established in literature. Koerner (2000) suggests that the applications that have been presented in Table 1.2 clearly demonstrate this aspect, and argues that geosynthetics are appropriate engineering materials along with concrete, steel and soil and that the unique characteristics of these materials only improve or complement the performance of civil engineering structures. He also indicates that quality control and quality assurance are essential aspects in the application of these materials. The practice of this technology has been making rapid strides, and international codes and regulations from North America and Europe and other countries are available for study. These can be modified to suit construction industry in India and other developing countries.

Bibliography

[1] Brandl, H. and Dalmatiner, J. (1986) *Struzmaners System 'NEW' und andere konstruktionen nack dem Baden-Aukerrabund-Prinzip*, Fed. Min. of Construction and Engineering Road Research, Vienna, 280.

[2] Fukuoka, M. (1980) Static and dynamic earth pressure on retaining walls, *Proc. 3rd Australian-New Zealand Conf. on Geomechanics*, Wellington, 3, 3-37, 3-46.

[3] Giroud, J.P. (1986) From geotextiles to geosynthetics, a revolution in geotechnical engineering, *Proc. 3rd International Conf. on Geotextiles*, Vienna, Vol. 1, pp. 1–18.

[4] John, N.W.N. (1987) *Geotextiles*, Chapman and Hall, New York.

[5] Jones, C.J.F.P (1996) *Earth Reinforcement and Soil Structures*, Thomas Telford Publishing, London.

[6] Jewell, R.A. (1996) *Soil Reinforcement with Geotextiles*, CIRIA, Thomas Telford, London.

[7] Koerner, R.M. (1998) *Designing with Geosynthetics*, 4th Edition, Prentice Hall, New Jersey.

[8] Koerner, R.M. (2000) Emerging and future developments of selected geosynthetic applications, *Journal of Geotechnical and Geo-environmental Applications*, ASCE, No. 4, 293–306.

2 Design Principles and Influencing Factors

2.1 Introduction

Reinforced soil structures are fundamentally different from conventional earth retaining systems which are externally stabilised in that they utilise a different mechanism for support and are internally stabilised. An externally stabilised system uses an external structural wall against which stabilising forces are mobilised, for example, a cantilever or gravity retaining walls and excavations supported with strutting, as shown in Fig. 2.1. An internally stabilised system involves reinforcements installed within and extending beyond the potential failure mass (tied back, reinforced soil walls and soil-nailed excavations). With this system, the interactions between the reinforcement and soil (to mobilise the tensile capacity of closely spaced reinforcing elements) eliminate the need for a structural wall or a support. Construction systems belonging to both the categories are shown in Fig. 2.1 (Jones 1996).

2.2 Mechanism of Reinforced Soil

The mechanism of reinforced soil can be explained in simple terms considering an element of cohesionless soil shown in Fig. 2.2(a). If a vertical stress is applied on the soil, it deforms both laterally (δ_h) and vertically (δ_v) and reaches a new equilibrium. If reinforcement in the form of plane sheet is introduced in the sample before the application of vertical stress on the sample, deformations are restrained due to the interaction between the soil and the reinforcement to some extent (δ_{hr} and δ_{vr}, Fig. 2.2(b)). Introduction of reinforcement generates inward lateral stress, which resists the shear stresses that are generated when a vertical stress (σ_1) is applied. If there is no significant deformation, the lateral stress is equal to K_0 times the vertical stress and this condition prevails at higher vertical stresses also. Assuming that the interaction behaviour between the soil and reinforcement is characterised by Mohr–Coulomb hypothesis, the changes in stress state using Mohr circle are indicated in Fig. 2.2(c). It can be noted that the Mohr circle is below the failure envelope and denotes safety. However, there are two conditions that can lead to failure. The shear stress at the interface of the soil and the reinforcement generates strains in the reinforcement, and tensile force is mobilised in the reinforcement. If the reinforcement force exceeds the tensile capacity of the sheet reinforcement, rupture failure occurs. Secondly, it is likely that a slip occurs between the soil and reinforcement if deformations are high or if the interface is smooth. These two conditions, viz. tensile failure and pullout failure, need to be examined to ensure the stability of the reinforced soil structures.

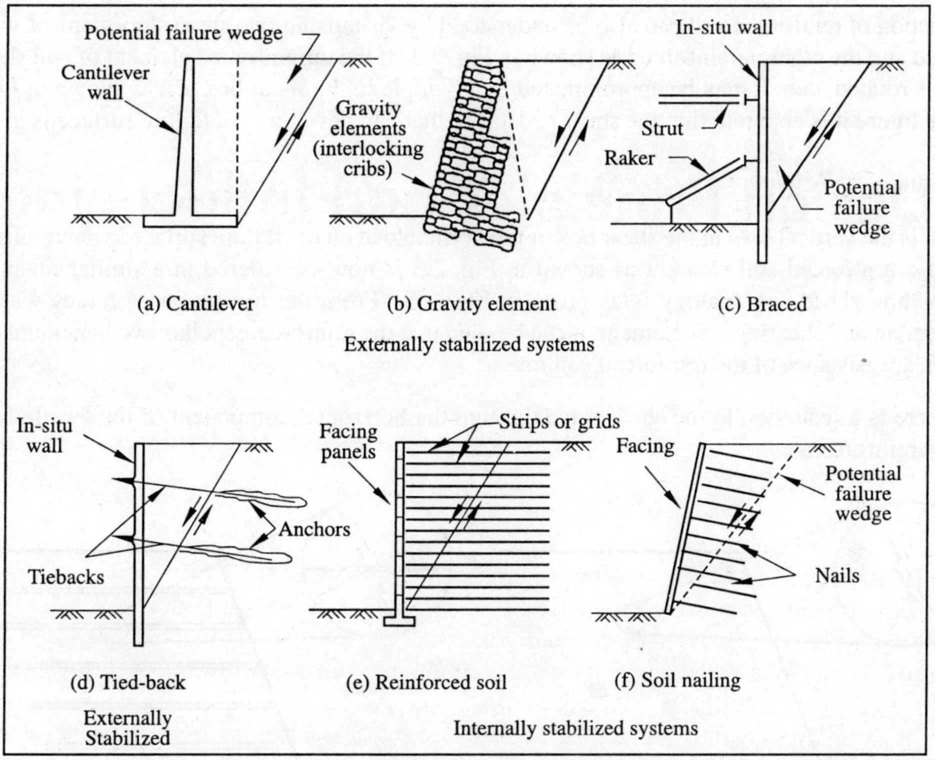

Fig. 2.1 *Internally and externally stabilized systems (Jones, 1996). (a–d) Externally stabilized; (e, f) internally stabilized.*

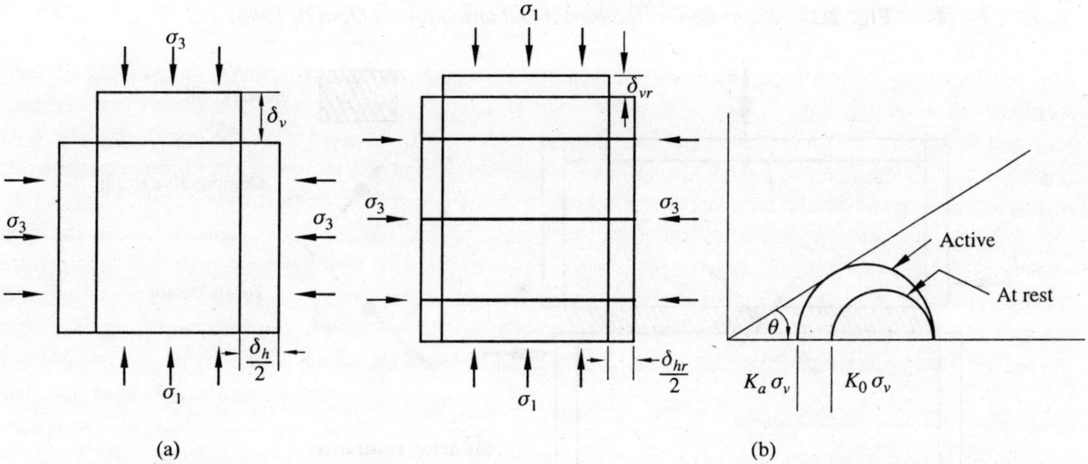

Fig. 2.2 *(a) Element of unreinforced soil, (b) element of reinforced soil, (c) state of stress in a reinforced soil.*

The action of reinforced soil can also be understood by visualising two steep slopes one of which is reinforced and the other unreinforced as shown in Fig. 2.3. If the unreinforced element of soil shown in Fig. 2.3 is rotated, then it may be approximated by a sample in the shear box test as shown in Fig. 2.4. From the figure it is apparent that the shear resistance that can develop on a failure surface is given by

$$P_{\text{resisting}} = P_v \tan\phi, \tag{2.1}$$

where P_v is the vertical load in the shear box or the normal load on the failure surface in an unreinforced slope. If a reinforced soil element as shown in Fig. 2.3 is now considered in a similar manner, the corresponding shear box analogy is as given in Fig. 2.5. From the resolution of forces within the reinforcement and shearing soil element, it can be seen that the reinforcement has two beneficial effects on the shear resistance of the reinforced soil mass.

- There is a reduction in the shear force through the horizontal component of the tensile force in the reinforcement.

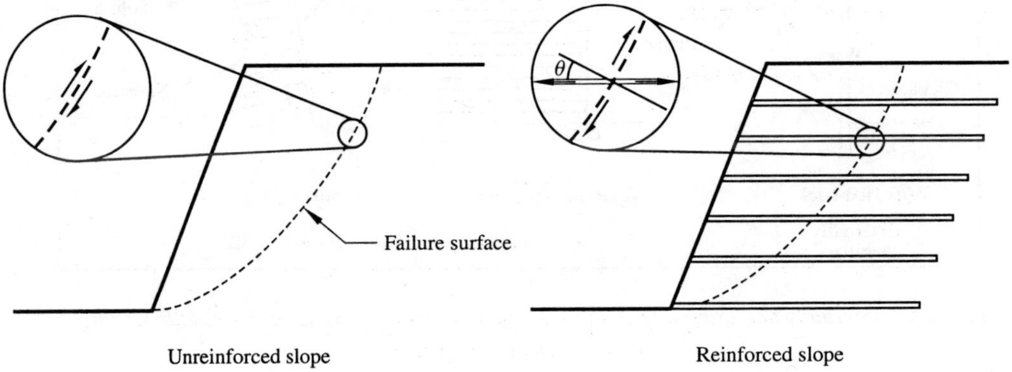

Fig. 2.3 *Steep slopes: reinforced and unreinforced (Jewell, 1996)*

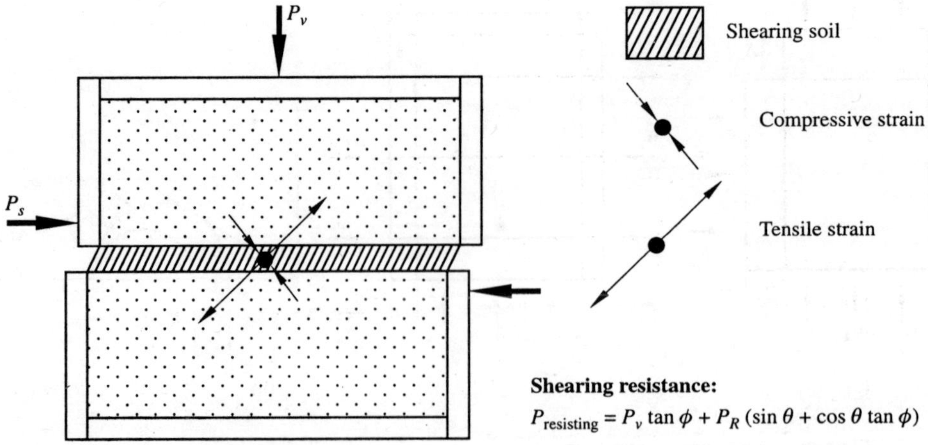

Fig. 2.4 *Idealised failure element in unreinforced slope (Jewell & Wroth, 1987)*

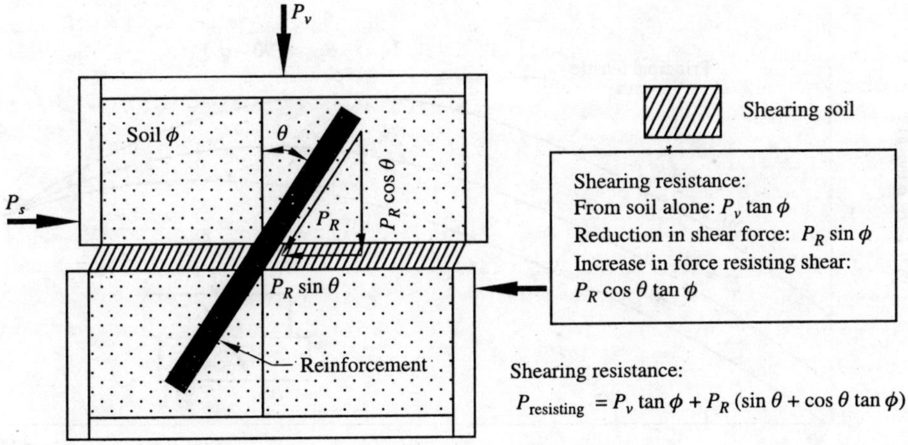

Fig. 2.5 *Idealised failure element in reinforced slope (Jewell & Wroth, 1987)*

- There is an increase in the normal force applied to the shear surface, and hence an associated increase in the shear resistance derived from the vertical component of the tensile force in the reinforcement.

The shearing resistance of the element of reinforced soil is increased from that given in Eq. (2.1) to that given in Eq. (2.2).

$$P_{\text{resisting}} = P_v \tan\phi + P_R(\sin\theta + \cos\theta \tan\phi). \tag{2.2}$$

Fundamental studies have shown that the reinforcement is most effective when aligned in the direction of tensile strain in the soil, so that a tensile reinforcement force develops (McGown et.al, 1978). The orientation (θ in Fig. 2.5) of the reinforcement with respect to the potential shear plane is the only geometry variable in Eq. (2.2). Hence it is apparent that the inclination of the reinforcement is the governing factor for determining how much additional shear resistance is generated by the inclusion of the reinforcement. Figure 2.6 demonstrates the variation of the function given in Eq. (2.2) with respect to different θ and ϕ angles. The optimum angle can also be obtained analytically by differentiating Eq. (2.2) with respect to θ, as follows.

$$\begin{aligned}
\frac{dP_R}{d\theta} &= P_R \cos\theta - P_R \sin\theta \tan\phi = 0 \\
&\Rightarrow \cos\theta = \sin\theta \tan\phi \\
&\Rightarrow \cot\theta = \tan\phi \\
&\Rightarrow \tan(90 - \theta) = \tan\phi \\
&\Rightarrow \text{Hence } \theta_{\text{opt}} = (90 - \phi).
\end{aligned} \tag{2.3}$$

It can be seen from Fig. 2.6 that the optimum orientation for the reinforcement is given by $\theta_{\text{opt}} = (90^\circ - \phi)$ (Jewell, 1996). In practice, the degree of improvement is critically dependent upon the magnitude of the mobilised reinforcement force (P_R). Two physical factors require consideration in the evaluation of P_R. The maximum force that the reinforcement can carry ultimately is governed

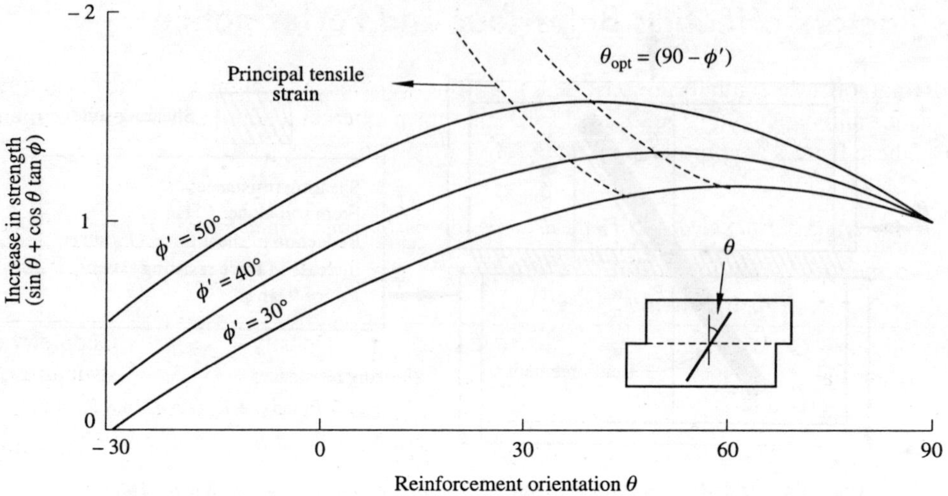

Fig. 2.6 *Shear resistance vs. reinforcement orientation (Jewell, 1996)*

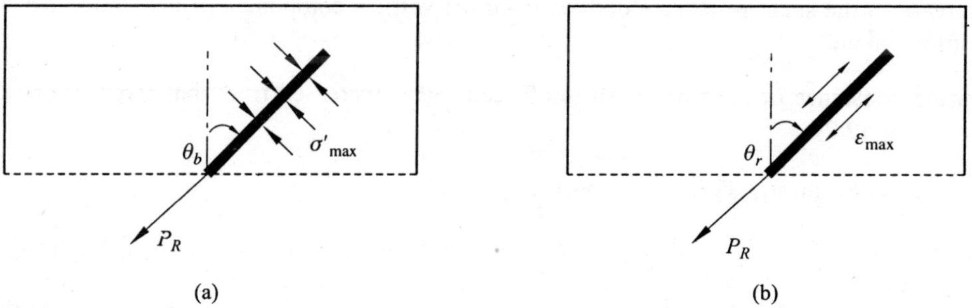

Fig. 2.7 *Optimum directions for reinforcement in soil (a) to maximise bond, (b) to maximise tensile strain (Jewell, 1996)*

by the bond between the soil and reinforcement. The bond is at a maximum when the maximum compressive stress in the soil acts perpendicular to the plane of the reinforcement. The stiffness of the reinforcement influences the soil shear deformation required to mobilise the reinforcement force. The maximum possible tensile strain in the reinforcement is equal to the tensile strain in the adjacent soil in the direction of the reinforcement. Thus, reinforcement oriented in the direction of maximum tensile strain will experience the greatest elongation for any given shear deformation in the soil. This optimum reinforcement angle (θ_r), is also shown in Fig. 2.7(a). Figure 2.7 shows the optimum directions (angles) to maximise bond (θ_b) and tensile strain (θ_r). These two optimum orientations are approximately equal (Jewell, 1996). In addition, in most situations such as bearing capacity improvement and slope stability improvement wherein the use of reinforcement is proposed, the inclination of the failure planes is in the range of 25° to 45° therefore and the provision of reinforcement becomes effective. The reader is referred to Jewell and Wroth (1987) for a more detailed discussion.

2.3 Factors affecting Behaviour and Performance

Reinforcing soil is essentially a construction technique. Hence, a number of factors influence the performance and behaviour of reinforced soil, apart form external loading and the environmental conditions. These factors are presented in Table 2.1.

Table 2.1 *Factors influencing the behaviour and performance of reinforced soil (Jones, 1996).*

Reinforcement	Reinforcement distribution	Soil	Soil state	Construction
Forms (fibre, grid, anchor, bar, strip)	Location	Particle size	Density	Geometry of structure
Surface properties	Orientation	Grading	Overburden	Compaction
Dimensions	Spacing	Mineral content	State of stress	Construction system
Strength		Index properties	Degree of saturation	Aesthetics
Stiffness				Durability

2.3.1 Reinforcement

The role of the reinforcement in the behaviour of reinforced soil is influenced by its form, surface properties, dimensions, strength and stiffness. A number of forms of reinforcement: strips, bars, grids/geotextiles/anchors and combined forms of reinforcement such as anchored grids, are available today. Forms such as steel strips, bars and geotextiles depend more on the interfacial friction between the soil and the reinforcement for mobilisation of tensile force in the reinforcement, whereas anchors depend on the bond or pullout resistance. Geogrids depend on the interfacial friction and bonding besides the interlocking effect, which develops when aggregates occupy the space in the apertures of the geogrids.

Surface properties play a crucial role in the mobilisation of friction and tensile forces in the reinforcement. The higher the frictional force between the soil and the reinforcement, more is the tensile force and hence more efficient is the effect of reinforcement in reinforced soil structure. Reinforcement with rough or roughened surfaces (sand glued to smooth surfaces, reinforcements embossed with some pattern of roughness such as grooves and cuts) perform better.

Dimensions of the reinforcement are arrived at after ensuring the strength and stiffness requirements from the design considerations. Depending on the availability of the material, the dimensions of reinforcement such as length and diameter/thickness can be optimised. Many of the reinforcing materials and geosynthetic materials are available in standard dimensions and length, and hence a proper choice assists considerably in reducing wastage of materials and thereby the costs.

Stiffness of the reinforcement has a significant influence on its performance. The longitudinal stiffness of the reinforcement governs the strain mobilization in the reinforced soil structure. The stress–strain characteristics of reinforcement are linear, whereas those of soil are nonlinear. For a stable condition the strain developed in the reinforced soil as a result of interaction between the soil and the reinforcement needs to be lesser than that mobilized in the soil alone. Assuming that tensile strain in the soil is the same as the strain in the soil (i.e., no slip), the maximum of reinforcement can be estimated using the theory of elasticity. It is useful to determine the magnitude of tensile strain that develops in the soil and the reinforcement to understand the equilibrium condition. This is useful for ensuring that: (a) the design values for the reinforcement and the soil resistance are realistically mobilized together, and (b) equilibrium is achieved with acceptable deformation of the structure (Jewell, 1996).

If the unreinforced soil (slope) has additional loading in the form of surcharge loading or earthquake or seepage forces, the driving forces (destabilising forces) increase. The factor of safety of slope (ratio of resisting forces to driving forces) is likely to decrease. When reinforcement is present, it becomes effective and tensile strains develop in the soil. As presented in the earlier section, the reinforcement force resists additional shear forces caused due to external conditions and increases the normal stresses, thereby eventually increasing the overall stability. In the reinforcement, the available reinforcement force increases as a function of the strain level. The rate of increase of available reinforcement force depends on the reinforcement properties. At equilibrium there is an optimum strain level which corresponds both to the soil and the reinforcement, and is often used for choosing the type of reinforcement depending on the function of the reinforcement. Equilibrium in a reinforced soil structure is obtained when the tensile strain in the reinforcement is equal to that in the surrounding soil. From a practical point of view the magnitude of the tensile strain possible within a reinforced soil structure is governed by two criteria:

- At the tensile strain at which equilibrium occurs the magnitude of the reinforcement tensile force and the shearing resistance of the soil can be realistically mobilised;
- The tensile strain at equilibrium can be achieved with acceptable deformations in the structure.

Hence, it is necessary to consider the stress–strain response of both the soil and the reinforcement individually and in a combined state. The stress–strain response of a typical granular material and normally consolidated clay under drained and undrained loading is demonstrated in Fig. 2.8. Figure 2.8(a) shows the response of loose and dense sands (drained) in a triaxial test. It can be noted that the deviatoric stress $(\sigma_1 - \sigma_3)$ increases and then decreases is in the case of dense sand as the axial strain is increased, whereas the deviatoric stress continuously increases in the case of loose sand. Figures 2.8(b) and 2.8(c) show the stress–strain response of normally consolidated clay in undrained and drained tests. Pore water pressure (u) developed in undrained test and volume change (ΔV) in drained set are also indicated. The stress–strain curves for typical reinforcing materials are given in Fig. 2.9(a). The tensile strain at equilibrium and the force in the reinforcing elements in a reinforced soil structure can thus be obtained theoretically from these two curves by combining them as shown in Fig. 2.9(b). The figure shows the stress available in the reinforcement from the conventional stress–strain curve (Fig. 2.9(a)) for the geosynthetic under consideration, and the stress required from the soil from the stress–strain curve of the soil (Fig. 2.8(a)). The relationship enables appreciation of the role of strains in reinforced soil behaviour, and is called strain compatibility curve. This compatibility curve is essentially a plot of: (a) the maximum required force for equilibrium in the soil which depends on the mobilised shearing resistance, and (b) the associated maximum available force from the reinforcement. From Fig. 2.9(b) it is apparent that:

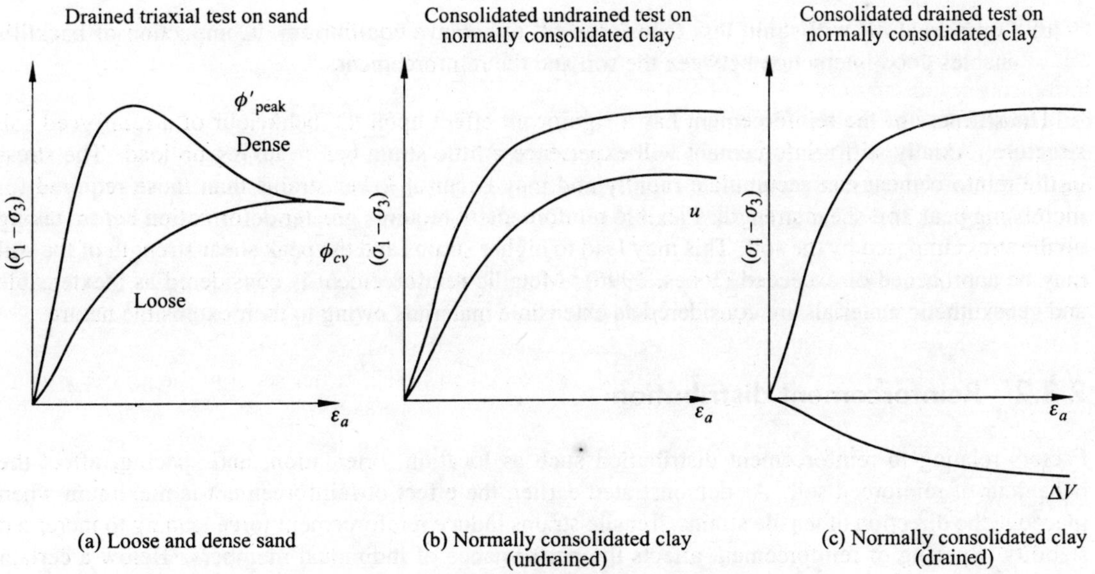

Fig. 2.8 *Stress–strain response of soils during triaxial test*

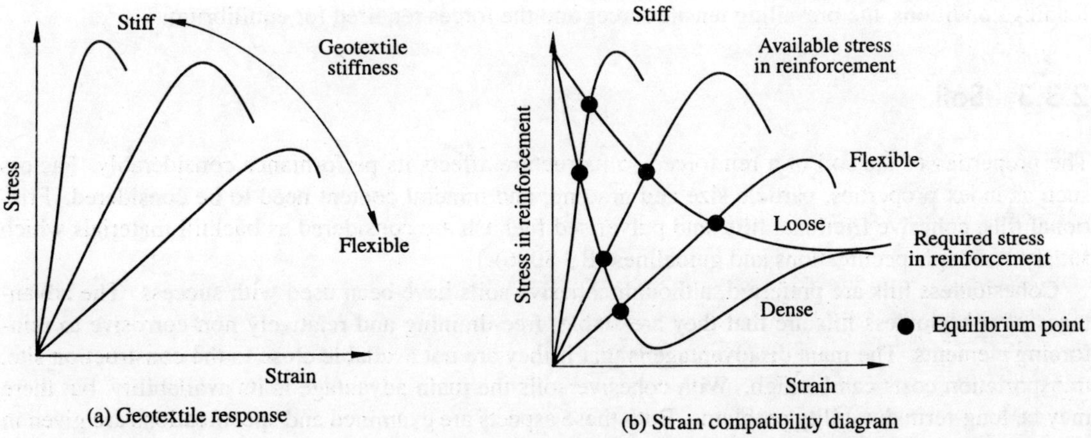

Fig. 2.9 *(a) Stress–strain response of geotextile reinforcement. (b) Strain compatibility diagram for equilibrium in reinforced soil (Jewell, 1996; Jones, 1996).*

(a) Stress in different materials (soil/reinforcement) is based upon strain level.

(b) Stiffness of reinforcement has a significant role. Stiffer/rigid reinforcements such as geogrids and steel can take higher stress; flexible reinforcing materials such as geotextiles will take less stress at the same strain levels.

(c) Compacted fill will strain less than loose fill to achieve equilibrium. Compaction of backfills enables good interaction between the soil and the reinforcement.

The stiffness of the reinforcement has a significant effect upon the behaviour of a reinforced soil structure. Axially stiff reinforcement will experience a little strain before taking up load. The stress in the reinforcement can accumulate rapidly and may occur at lower strains than those required for mobilising peak soil shear strength. Flexible reinforcement requires greater deformation before taking up the stress imposed by the soil. This may lead to higher strains and the peak shear strength of the soil may be approached or exceeded (Jones, 1996). Metallic reinforcement is considered as inextensible and geosynthetic materials are considered as extensible materials owing to their extensible nature.

2.3.2 Reinforcement distribution

Factors relating to reinforcement distribution such as location, orientation, and spacing, affect the behaviour of reinforced soil. As demonstrated earlier, the effect of reinforcement is maximum when placed in the direction of tensile strains. Tensile strains induce reinforcement force leading to increased stability. Spacing of reinforcement affects the performance of individual members. Below a certain spacing interference occurs, and it is not likely to be effective as the increase in shear strength is less. However, in practice, the choice of spacing depends on other factors such as compaction, lift thickness, availability of the materials of required specification, etc. It will be demonstrated in a subsequent chapters that the optimum spacing can be arrived at after considering the type of structure, external loading conditions, the prevailing tensile forces and the forces required for equilibrium.

2.3.3 Soil

The properties of the soil of a reinforced soil structure affects its performance considerably. Factors such as index properties, particle size and grading, and mineral content need to be considered. Frictional fills, cohesive frictional fills, and pulverised fuel ash are considered as backfill materials which satisfy standard specifications and guidelines (BS 8006).

Cohesionless fills are preferred, although cohesive soils have been used with success. The advantages of cohesionless fills are that they are stable, free-draining and relatively non-corrosive to reinforcing elements. The main disadvantage is that if they are not available close to the construction site, transportation costs can be high. With cohesive soils the main advantage is its availability, but there may be long-term durability problems. Both these aspects are examined and specifications are given in the code of practice (BS 8006).

Cohesionless fills

Cohesionless fill is made of a good quality well-graded non-corrosive material usually specified by the angle of internal friction. Knowledge of the following material properties is required for the selection of cohesionless fill:

(a) density
(b) grading
(c) uniformity coefficient, C_u
(d) pH value

(e) chloride ion content
(f) SO_3 content
(g) resistivity
(h) redox potential
(i) angle of internal friction under effective stress conditions, ϕ'
(j) coefficient of friction between the fill and the reinforcing element, μ

The density of the backfill material has a direct effect upon its internal and external stability. The soil should be well graded and have the requirements given in Table 2.2.

Table 2.2 *Grading requirements for frictional fill*

Sieve size	% passing
125	100
90	85 – 100
10	25 – 100
600 μm	10 – 65
63 μm	0 – 10
2 μm	0 – 10

The uniformity coefficient, C_u, is the ratio of the maximum size of 60% of the sample to the effective size (size passing 10%). A value of $C_u \geq 5$ is considered acceptable. Properties such as pH value, chloride ion, SO_3 content, resistivity and redox potential are associated with the durability of the reinforcing materials. Acceptable values for different reinforcing materials are given in Table 2.3. The effective angle of internal friction of cohesionless soil (ϕ') should be a minimum of $\geq 25°$. Friction between the fill and the reinforcing elements is an important property for characterizing the interaction between the soil and the reinforcements. This is usually measured from direct shear tests.

Table 2.3 *Acceptable limits for reinforcing materials (Jones, 1996)*

Reinforcing materials	pH value		Max. chloride ion content %	Total SO_3 %	Min. resistivity (ohm-cm)	Minimum redox potential (volts)		Max. microbial activity index	Max. organic content %
	min.	max.				Frictional fill	Cohesive frictional fill		
Aluminium alloy	6	8	0.05	0.5	3000	0.4	0.43	5	0.2
Copper	5	9	0.05	0.5	2000	0.25	0.25	5	0.28
Hot-dip	5	10	0.02*	0.1*	1000*	–	0.43	5	0.2
Galvanized steel	–	–	–	–	–	0.4	–	–	–
Stainless steel	5	10	0.025*	0.1*	3000	0.35	0.35	5	0.2
Fibretain (FRP)	4	9	2.0	1.0	1000	NA	NA	NA	NA
Paraweb	–	–	–	–	–	NA	NA	NA	NA
Tensar	–	–	–	–	–	NA	NA	NA	NA

* When the structure is submerged in fresh water, these values are reduced.

Cohesive frictional fill

The main advantage of cohesive frictional fill is its better availability when compared with frictional fill. Cohesive frictional fill is specified in standards such as the UK code of practice for reinforced soil, BS 8006:2000. Knowledge of the material properties such as:

(a) cohesion under effective stress conditions,
(b) adhesion between the fill and the reinforcing elements under effective stress conditions,
(c) liquid limit (LL),
(d) plasticity index (PI), and
(e) consolidation parameters

is required for the selection of cohesive frictional fill. Besides these properties, requirements for grading such as density and friction angle need to be satisfied for the case of cohesionless materials (Table 2.4).

Table 2.4 *Grading requirements for cohesive frictional fill*

Sieve size	% passing
125	100
90	85 – 100
10	25 – 100
600 μm	11 – 100
63 μm	11 – 100
2 μm	0 – 10

Similar to the considerations presented for cohesionless soils, the coefficient of uniformity $C_u \geq 5$ and angle of internal friction $\phi' \geq 20°$ are expected. For materials finer than 52 μm liquid limit, LL $\leq$ 45% and plasticity index PI $\leq$ 20% are considered acceptable. A limit of 10% clay is related to the need for the fill to be sufficiently free-draining to provide stability during construction. The clay content of soil can be difficult to measure and large variations are common. The liquid limit and the plasticity index may give a more direct indication of mechanical properties, in which case one need not attempt to determine the clay content. The durability criteria (pH value, chloride ion, SO_3content, resistivity and redox potential) associated with cohesive frictional fill are the same as for frictional fill (Table 2.3). Moisture content in the range close to optimum will normally produce satisfactory conditions with respect to stability and handling ability.

Waste materials

The use of waste materials as fill for reinforced soil structures is attractive from an environmental as well as economic viewpoint. Waste products such as mine waste and pulverised fuel ash (PFA) that are produced in abundant quantities can be used as backfills provided their use is approved after thorough testing. In India, both bottom ash and fly ash have been successfully used as backfill materials in reinforced soil retaining wall construction projects by the Central Road Research Institute, New Delhi.

2.3.4 Soil state

Factors such as density overburden, state of stress, drainage conditions and degree of a saturation, affect the response of reinforced soil structure. In drained conditions dense soil dilates during shear, whereas loose soil shows volume reduction. In undrained state, negative pore water pressure develops in dense soil, which contributes to the increase of normal stresses on the reinforcement. Figure 2.8 shows typical stress–strain response of soils during shear in a triaxial test under drained and undrained conditions. Bolton (1986) proposed a simplified relationship between the peak friction angle (ϕ_p') and the residual or constant volume friction angle (ϕ_{cv}) given by

$$\phi_p' = \phi_{cv} + 0.8\psi,$$

where ψ is the angle of dilatancy. It denotes the effect of compaction and the tendency for dilation. Higher angles of dilation denote higher normal stress on the reinforcement, which increases the pull-out resistance of the reinforcement.

The state of stress in the reinforced soil needs to be assessed to know the earth pressure distribution behind the retaining wall. Normally, when stiff reinforcements are provided, at rest conditions prevail in retaining walls of height less than 6 m. Some of these aspects pertaining to reinforced earth retaining wall design are discussed in Chapter 6. If clayey soils are used as backfill materials, reinforced soil behaviour at low degrees of saturation impede the flow movement, and as a consequence transfer of effective stresses is slow.

2.3.5 Construction factors

Construction variables such as geometry of the structure, compaction, construction system, aesthetics and the durability of the reinforcing elements affect the behaviour of reinforced soil structures. It is possible to have different facing systems for the vertical retaining wall, which may lead to variations in earth pressure distributions behind the retaining wall.

Compaction in the field induces considerable lateral earth pressure. Shear strains are induced in the soil as a result of compaction and the reinforcing members will resist the shear strain. It is necessary to estimate the effect of compaction pressures and account for them so that the reinforcement members at the level where the compaction stresses are predominant do not fail. Compaction is normally done using rollers parallel to the face of the wall. If the influence of stresses is likely to be more, static compaction can be used.

Reinforcing soil is essentially a construction technique and the behaviour of a reinforced soil structure depends on the construction systems/methods adopted. Construction methods are identified with the use of certain types of reinforcement forms, materials, facing elements and connection elements. These techniques influence the earth pressure distribution, the tensile forces mobilised in the reinforcement and the deformation.

2.4 Soil–Reinforcement Interactions

Besides the intrinsic properties of the soil and the reinforcement, the interaction between the soil and the reinforcement is an essential consideration for determining the reinforced soil behaviour. The rate of change of axial force in the reinforcement embedded in the soil is controlled by the bond or the pull-out resistance. For the reinforcement in a slope or a retaining wall, the changes in axial force close to the free end of the reinforcement are important for determining the necessary lengths to enable the required axial forces to be generated. The pull-out resistance is due to friction, adhesion or bearing stresses between the soil and the reinforcement, but the relative movement of the soil and the reinforcement is different from the movement in direct sliding.

For soil on either side of strip or sheet reinforcement, the interaction between the soil and the reinforcement is the same as in direct sliding and bond. The only difference is that in the reinforced case the normal effective stress on the reinforcement may increase above the level of the mean overburden stress as a result of the local dilation of dense granular soil around the reinforcing element. With grid reinforcement, the bond is induced by the shear along the surfaces of the reinforcing element and by the bearing of the transverse or anchor members of the grid against the soil. No soil-to-soil friction is developed as there is no relative movement of the soil on either side of the grid. This situation cannot be reproduced in a shear box test and the bond for grids may only be measured in pull-out tests in which the reinforcement is pulled out of the stationary soil. Considerable care has to be taken to ensure that the pull-out force is not unduly influenced by the boundary conditions of the test.

2.5 Design of Spacing and Layout

Walls and abutment structures together with reinforced slopes are conventionally constructed using horizontal reinforcement. The vertical spacing of the reinforcement may remain constant throughout the full height of the structure, but the density or strength of the reinforcement is likely to be greater towards the base where the horizontal thrusts are the maximum. The simplest layout of reinforcement is a uniform distribution of identical reinforcing elements throughout the entire length and height of the structure (Jones, 1996). This layout is usually uneconomical except for small-scale works. A more economical design may be achieved by using reinforcement of different properties, or by dividing the structure into different zones.

Bibliography

[1] Bolton, M.D. (1986) The strength and dilatancy of sands, *Geotechnique*, 36, No. 1, 65–78.

[2] BS 8006 (1995) British Standard: Code of practice for strengthened/reinforced soils and other fills.

[3] Jewell, R.A. (1996) *Soil Reinforcement with Geotextiles*, Construction Industry Research and Information Association (CIRIA) Special Publication 123.

[4] Jewell, R.A. & Wroth, C.P. (1987) Direct shear tests on reinforced sand, *Geotechnique*, Vol. 37, No. 1, pp 53–68.

[5] Jones, C.J.F.P. (1996) *Earth Reinforcement and Soil Structures*, Thomas Telford Publishing, London.

[6] McGown, A., Andrawes, K.Z. & Al-Hasani, M.M. (1978) Effect of inclusion properties on the behaviour of sand, *Geotechnique*, Vol. 28, No. 3, pp 327–346.

3 Materials and Material Properties

3.1 Introduction

Knowledge of materials and their properties is essential for the design and analysis of structures constructed with reinforcement/geosynthetics. The material properties related to the behaviour, manufacture and quality control are generally referred to as index properties and those related to the design as design or performance properties (Koerner, 1998). For example, in constructions dealing with soils, liquid limit, plastic limit, etc. are called the index properties and compression index, permeability coefficient and shear strength are called the design/performance properties. Similarly, both index and design/performance properties of geosynthetics and soil reinforcement are relevant for specifying materials for use in civil engineering applications. Rao (1990) indicates that these properties are helpful in:

(a) determining the suitability of a material for specific application/function,
(b) quality control of a product during production and installation,
(c) comparison of products for selection of the appropriate one, and
(d) specification of products for specific applications.

3.2 Materials

A wide variety of synthetic or processed and natural materials ranging from metals, polymers, jute, and coir are being used today in different applications. Considerable information exists on these materials and testing procedures (Sharma and Lewis, 1994; Koerner, 1998). The following sections provide some details on metallic reinforcement and geosynthetics; the treatment however is not exhaustive. Good information on the properties of natural materials such as coir (Rao and Balan, 2000) and jute (Rao, 1996) is available in literature.

3.2.1 Metallic reinforcement

Materials such as steel in the form of sheets, grids, meshes, strips, bars, rods (Fig. 3.1) that are capable of sustaining tensile loads and restraining deformations have been used for many years. Steel has been used successfully for a long time in soil for piling and pipelines. For using these materials as soil reinforcement, information with regard to their tensile strength behaviour, soil-reinforcement friction and durability behaviour is necessary. Properties such as tensile strength, Young's modulus,

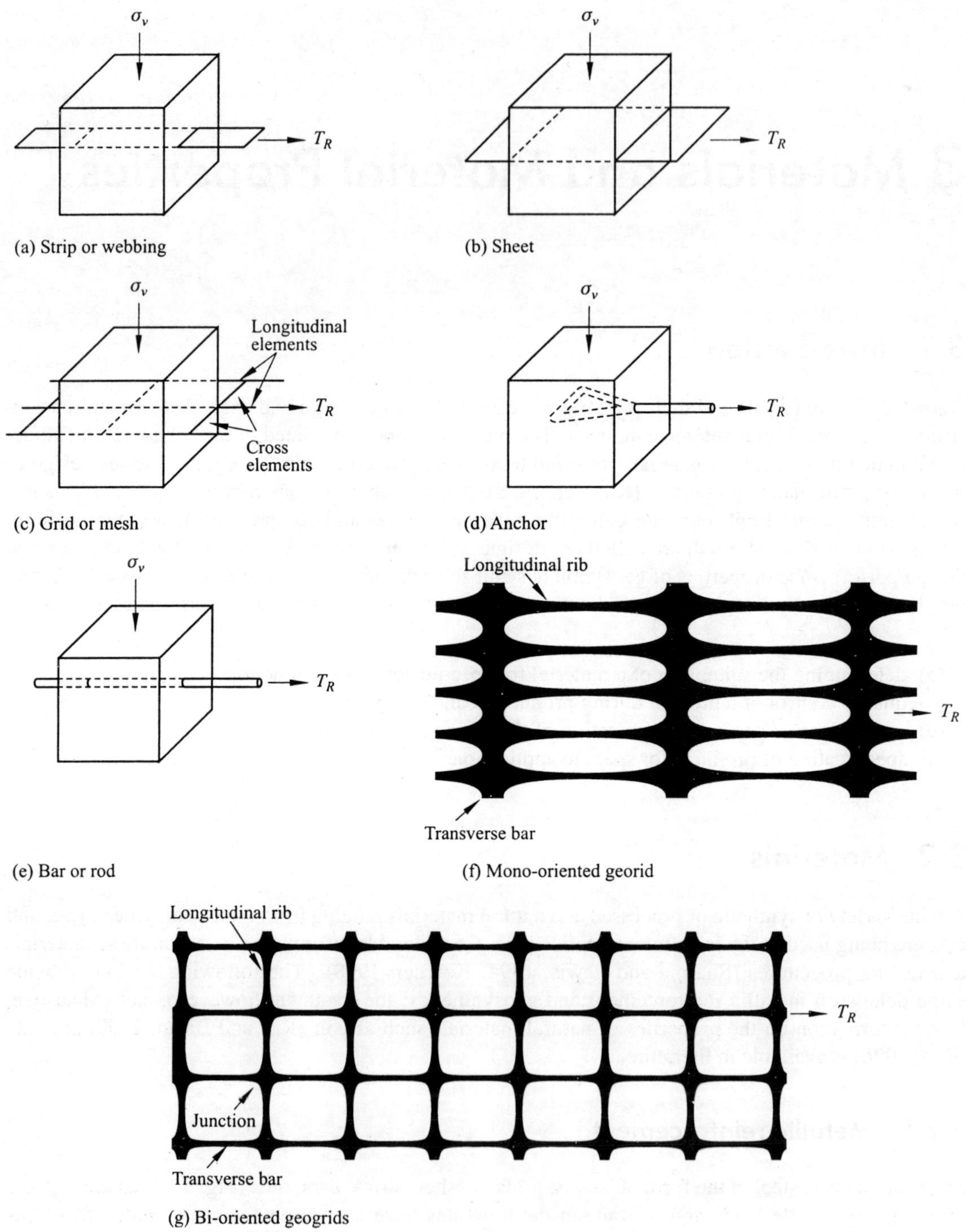

Fig. 3.1 *Various types of reinforcement members*

and stress–strain behaviour can be obtained by conducting tension tests on reinforcing materials. The interface properties of soil–reinforcement can be obtained from direct shear tests. Based on the long term degradation characteristics of steel, BS 8006: 1995 recommends different types of steel that can be used as soil reinforcement (Table 3.1). The code also gives guidelines on the sacrificial thickness that needs to be provided depending on the design service life and the environment. The details are given in Table 3.2. It may be noted that different types of steel identified for use are as per the British Standards (BS) and European Standards or Norms (EN). Materials with equivalent properties may be used in other countries.

Table 3.1 *Minimum properties of some different types of steel reinforcement (BS 8006)*

Type of steel reinforcement	Maximum thickness (mm)	Tensile strength σ_t (N/mm^2)	Shear strength σ_q (N/mm^2)	Bearing strength σ_{bc} (N/mm^2)
Carbon steel (BS EN 10025 : 1993 S 235 JR)	16	340	205	340
Carbon steel (BS EN 10025 : 1993 S 355 JR)	16	490	295	490
Stainless steel (BS 1449:Part 2:1993 316S31, 316S33)	10	510	305	510
Stainless steel (BS 1449:Part 2:1983 316S31 CR temper rolled to a minimum 0.2% proof stress of 310 N/mm^2)	6	650	390	650
Carbon steel rod (BS 4449 : 1988)	40 diameter	485	290	485

3.2.2 Polymeric materials

Majority of geosynthetics is of polymeric origin, composed of polypropylene, polyethylene or polyester fibers; a small percentage is of polyamide orgin. The types of fibers used in the manufacture of geotextiles include monofilament, staple, and slit film. If fibers are twisted or spun together, they are known as yarn. Monofilament fibers are created by extruding molten polymer through an apparatus known as spinnaret, containing several small-diameter holes. The extruded polymer strings are then cooled and stretched to align the polymers and give the fiber increased strength. The staple fibers are then spun into longer fibers known as staple yarns. Slit-film fibers are manufactured by extruding a continuous sheet of polymer and cutting them into fibers by knives or air jets.

The fibers or yarns are formed into geotextiles using either woven or non-woven (spun-bonded) methods. Woven geotextiles are formed using traditional weaving methods and weave types. Common terminology associated with woven geotextiles include machine direction, cross-machine direction, warp, and weft. The machine direction refers to the direction in the plane of fabric parallel to the direction of manufacture, and conversely, the cross-machine direction refers to the direction in the plane of fabric perpendicular to the direction of manufacture. The machine direction is also known as

Table 3.2 *Sacrificial thickness to be allowed on each surface exposed to corrosion (BS 8006)*

Design service life (years)	Reinforcement material	Sacrificial thickness (mm)	
		Land-based structure (out of water)	Freshwater structure
5	B	0.25	0.25
	G	0	0
	S	0	0
10	B	0.35	0.40
	G	0	0
	S	0	0
50	B	1.15	1.55
	G	0.30	0.55
	S	0.05	0.07
60	B	0.38	1.68
	G	0.05	0.63
	S	0.05	0.09
70	G	0.45	0.07
	S	0.05	0.10
120	G	0.75	1.0
	S	0.10	0.2

Notes:
1. B: Black steel (ungalvanized); G: galvanized steel; S: stainless steel.
2. Black steel should not be used as a reinforcement material for a design service life greater than 60 years.
3. Linear interpolation may be used for intermediate service lives.
4. These values apply to steels embedded in fills of different classes in the specification for highway works in UK.
5. Sites of special aggressiveness are to be assessed by specific study.

the warp since warp yarns are those that are placed lengthwise on the weaving loom; and the cross-machine direction is known as the weft since weft yarns are woven in-between and perpendicular to the warp yarns.

To obtain non-woven geotextiles, the fibers are placed and oriented on a moving conveyor belt. The fibers are bonded by needle punching, melt bonding, or resin bonding. The needle punching process consists of pushing numerous barbed needles through the fiber web. The fibers are thus mechanically interlocked into a stable configuration. The melt bonding process consists of melting and pressurizing fibers together at their crossover points. In resin bonding, an acrylic resin is applied to the fiber web to form the geotextile. In the following sections, a brief review and evaluation of properties of geotextiles is presented.

Polymers are essentially a chain of molecular compounds having different molecular weights as well as specific distribution of molecular weights. The molecular weights and their distribution influence the mechanical properties of the geosynthetic material. Hence the knowledge of raw polymer properties is useful.

The resin density of a polymer is related directly to the crystallinity of the polymer and thus to the behaviour of the geosynthetic. The density of geosynthetic material is usually given as the resin or raw polymer density. The range of densities are usually similar but may differ from the manufactured sheet density due to additives used during the manufacture of the sheet; it will also differ if the sheet is manufactured from more than one polymer. The density of most polymers falls between 0.85 and 1.50 g cm^{-3}.

The melt index test is an important test conducted during manufacturing. The test determines the fluidity of the molten geosynthetic resin by using a constant load to push it through a standard orifice. The melt index is taken as the weight of molten resin pushed though the orifice in 10 minutes. The value is typically given in units of grams per 10 minutes and for polyethylene ranges from 0.2 to 1 g/10 min. Melt index is directly related to the mean molecular weight and gives an indication of the necessary processing parameters. High melt index values indicate a low-molecular-weight polymer, and vice versa.

To understand the behaviour of geosynthetics better, it is sometimes important to identify them clearly on a molecular basis. Several chemical identification methods are available to characterize polymeric materials. These methods are generally referred to as *fingerprinting* and include:

- Thermogravimetric analysis (TGA)
- Differential scanning calorimetry (DSC)
- Thermomechanical analysis (TMA)
- Infrared spectroscopy (IR)
- Gas chromatography (GC)
- Gel permeation chromatography (GPC)

TGA, DSC and TMA fingerprint geosynthetics using thermal methods. TGA records the mass of a polymer as a function of temperature. Generally, as the temperature increases the mass of the specimen decreases due to the vapourization of specific components. DSC is performed by monitoring the heat flow into and out of a specimen placed in an increasing temperature environment. The heat flow is measured by maintaining a temperature balance between a reference cell and the test specimen cell. The curves generated from DSC depict the crystallization process of a polymer.

The TMA test measures the penetration, expansion, contraction, or extension of the geosynthetic material as a function of temperature. The test is relatively straightforward and provides information on the coefficient of linear expansion, softening point, and the glass transition temperature. The IR and GC tests can be used to identify additives and plasticizers. IR identifies the surface chemistry of polymeric materials by scanning them using a range of infrared frequencies and correlating the absorption to specific atomic groups. The GC test requires vapourization of volatiles within the specimen in a confined headspace. The resulting gas is analyzed to identify the various components present in the specimen.

For an accurate determination of the molecular weight distribution of a specimen, a GPC test should be performed. The test is performed by passing the polymer (in solution form) through a rigid porous gel column, which acts as a sieve. The higher-molecular-weight polymers pass through the column more quickly than the lower-molecular-weight polymers; this allows the molecular distribution to be analyzed.

Geotextiles

The material properties generally specified are thickness, mass per unit area, uniaxial tensile strength, multiaxial tensile strength, puncture resistance, trapezoid tear strength, apparent opening size, permittivity, transmissivity, and ultraviolet resistance.

Thickness

The average thickness of a geotextile is measured using a thickness gauge under a gradually applied specified pressure. For geotextiles, a pressure of approximately 2 kPa is typically used. The thickness of a geotextile is related to other material properties such as mass per unit area, tensile strength, puncture resistance, and tear resistance.

Mass per unit area

The mass per unit area of a geotextile is determined by weighing several test specimens of known area, taken from various locations of the fabric sample. The calculated values are averaged to obtain the mean mass per unit area of the sample. Many specifications attribute a certain mass per unit area to a certain set of mechanical and hydraulic properties, such as puncture resistance, tear resistance, apparent opening size and tensile strength.

Uniaxial tensile strength

The uniaxial tensile strength of a geotextile is measured in a tensile testing machine by applying a continually increasing load along the longitudinal length of a specimen. The specimen is held within clamp, specially designed to prevent slippage. While the test values typically reported are the breaking load and apparent elongation (reported as a percentage increase in length), a load elongation curve is generated by dividing the load by the width and thickness of the geotextile specimen. Since the thickness of the geotextile typically decreases as a tensile load is applied and is also variable throughout the specimen, the value is often reported as the load divided by the specimen width (kN/m). Researchers have studied the factors affecting the uniaxial tensile strength of geotextiles; this is available in the literature on geosynthetics. These factors include specimen size, aspect ratio (width-to-length ratio), strain rates, gauge length, clamping conditions, fabric type, field behaviour and anisotropic conditions.

Standard test methods have been developed for uniaxial geotextile tensile testing. The two commonly used standards include the grab (ASTM D 4632) and wide width (ASTM D 4595) methods. Figure 3.2 shows various tensile test specimen sizes. The strip and grab tensile tests utilize procedures originally established for the textile industry. As the tensile load is applied to a specimen, the specimen necks in its central region. In the grab tensile test, the clamps holding the specimen do not hold the entire width of the specimen. The grab method measures the effective strength of the geotextile, that is, the strength of the material in a specific width, together with the additional strength contributed by adjacent material. Both grab and strip tests are useful as quality control and acceptance tests. The recommended tensile test for design is the wide-width tensile test, ASTM D 4595. This test was developed specifically for geotextiles and uses an 200 mm wide sample. The increased specimen width minimizes the necking or contraction effects that occur in the strip tensile test and also provides a closer model of the behaviour expected in the field. The major concern with the wide-width test is the slippage of the specimen from the jaws. Also, if the specimen is held too tightly in the jaws, the jaws may damage the geotextiles. ASTM D 4595 allows the substitution of a 100 mm wide specimen due to clamping and equipment limitations.

Burst tests

Burst tests simulate three-dimensional loading conditions for geotextiles under wheel loads, puncture situations, in settlement or subsidence areas. The test commonly uses a diaphragm that is expanded by fluid pressure until the geotextile specimen ruptures. The bursting strength is the difference between the total pressure required to rupture the specimen and the pressure required to inflate the diaphragm.

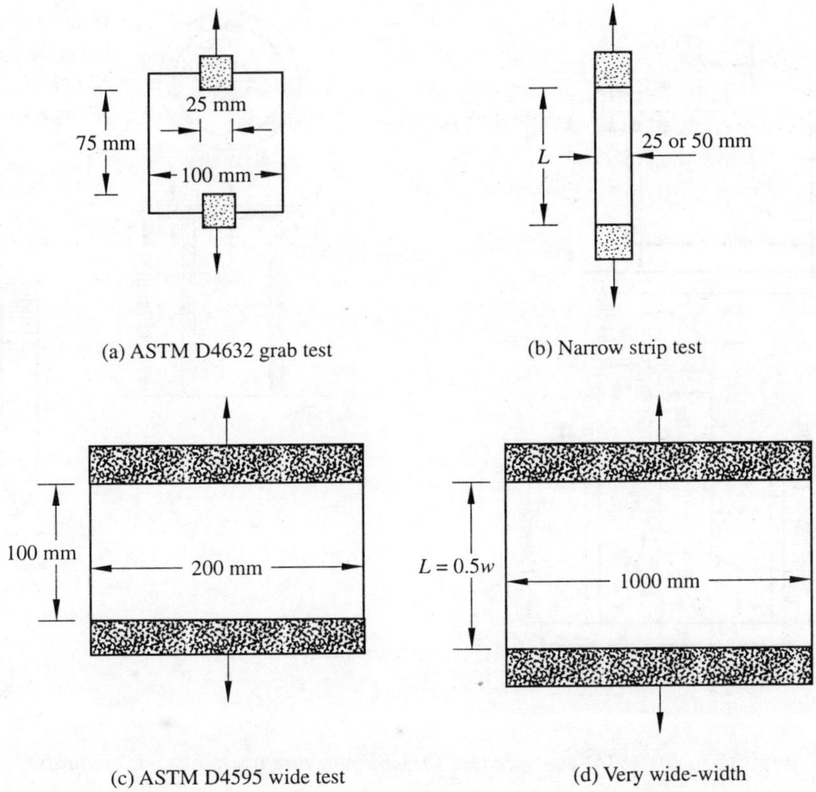

Fig. 3.2 *Various tensile test specimen sizes: (a) ASTM D4632 grab test, (b) narrow strip test, (c) ASTM D4595 wide-width, (d) very wide-width test (Koerner, 1998)*

The puncture resistance of a geotextile is determined by measuring the force required by a probe to punch through the geotextile. Two commonly used methods to measure puncture resistance are the ASTM D 4833 and the CBR puncture methods. The geotextile is held without tension in a ring clamp attachment consisting of concentric plates with an open internal diameter of approximately 46 mm. The probe is pushed through the center of the geotextile at constant rate and the maximum force required to puncture the geotextile is recorded on the test apparatus. The details are shown in Fig. 3.3(a). The CBR puncture resistance test is a standard test in Germany. The test apparatus is illustrated in Fig. 3.3(b) and consists of a 250 mm test specimen clamped without tension between two plates and attached to a California Bearing Ratio (CBR) mold. A 50 mm diameter center-steel plunger having a flat end and 45° chamfered edge is pushed through the geotextile at a constant rate. The maximum force required to puncture the geotextile and the maximum elongation are recorded.

Trapezoid tear strength

The tear strength tests measure the ability of a geotextile to resist tear once a tear is initiated. In the trapezoidal tear test, an isosceles trapezoid shape is marked on a rectangular specimen and a 16 mm long preliminary cut is made at the center of a 25 mm edge. The maximum recorded force causing tearing of the specimen is known as the trapezoidal tear strength.

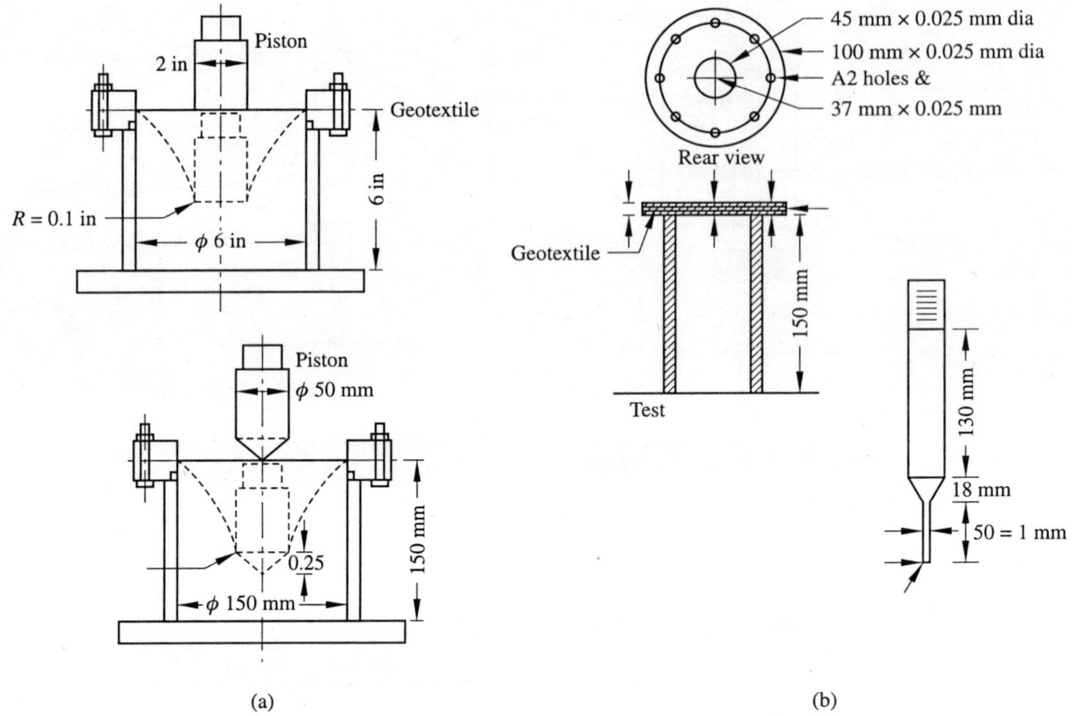

Fig. 3.3 *(a) ASTM puncture test, (b) CBR puncture resistance test apparatus*

Soil geosynthetic friction tests

The friction or shear strength between soils and the geosynthetics and between the geosynthetics and geosynthetics, is a critical design parameter in many applications. The friction values are important because the combination of soils adjacent to the geosynthetics or the geosynthetics adjacent to the geosynthetics may form a weak plane on which a preferential sliding plane may develop. The tests use conventional soil direct shear testing devices or a modification of the devices for the pull-out to measure these parameters. Figure 3.4 demonstrates the use of direct shear test apparatus to evaluate interfacial friction angle between the soil and the geosynthetic material.

Apparent opening size

The pore size of a geotextile may be determined by direct or indirect methods. The direct methods involve image analyzers. Indirect methods are most commonly used and include dry sieving, wet sieving or mercury intrusion. Of these, the dry and wet sieves are more popular. The apparent opening size (AOS or O_{95}) is a dry sieve method used to estimate the approximate largest particle that would effectively pass through a geotextile. To perform the test in accordance with ASTM D 4751, a geotextile specimen is placed in a sieve frame, and sized glass beads ranging from 0.075 to 2.0 mm are successively sieved through the specimen. The AOS is assigned as the bead size designation for which 5 percent or less (by weight) of the beads pass the geotextile specimen, and is reported in millimeters. The wet sieve methods add a water spray to force particles dynamically through the geotextile. The AOS determined using hydrodynamic methods is known as the filtration opening size (FOS).

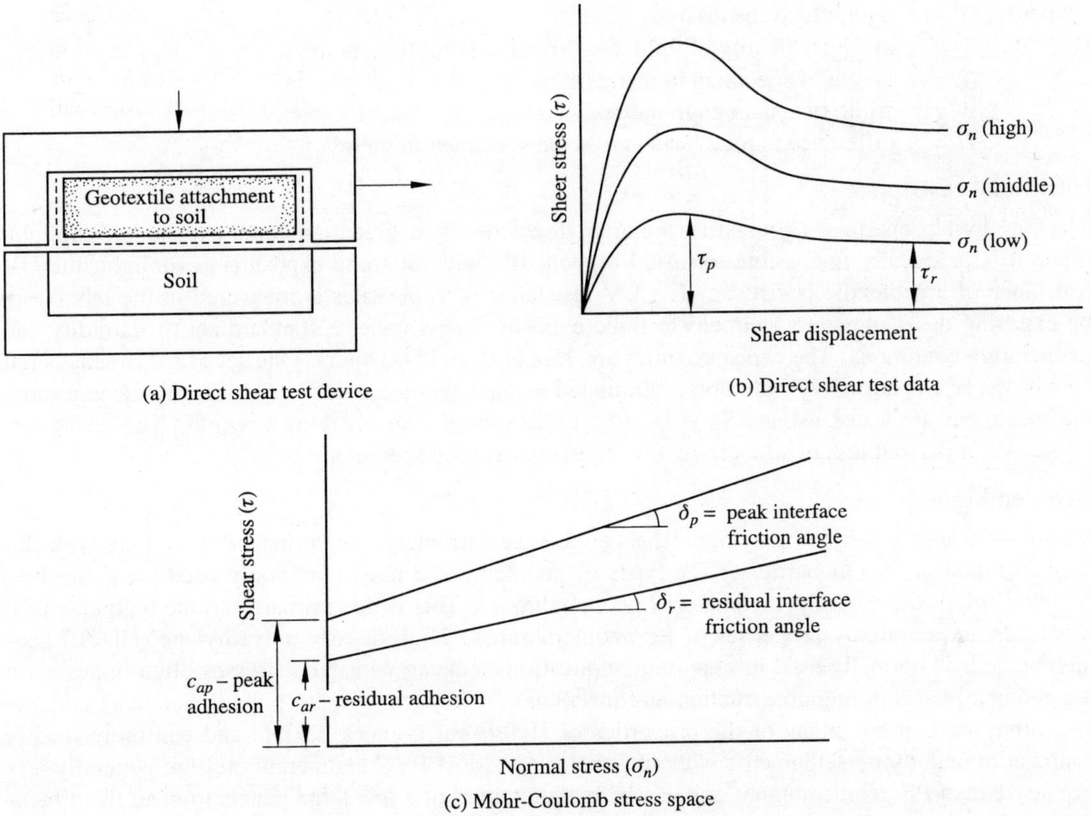

Fig. 3.4 *Typical experimental setup and test results to determine interfacial friction parameters between soil and geosynthetic materials*

Permittivity

Permittivity is a measure of the cross-plane or perpendicular flow of water through a geotextile, which may be used in designing geotextiles for filtration or drainage applications. It is measured using constant or falling head tests, similar to those used to determine the coefficient of permeability of soils. Permittivity values are reported in terms of coefficient of permeability divided by the geotextile thickness. The unit for permittivity is 'per second' (s^{-1}).

Transmissivity

Transmissivity is the measure of the in-plane flow of water through a geotextile. ASTM D 4716 measures hydraulic transmissivity by measuring the quantity of water that passes through a saturated test specimen in a specific interval under a specific normal stress and hydraulic gradient. The hydraulic transmissivity is calculated by the equation

$$\theta = QL/WH, \tag{3.1}$$

where $\theta =$ hydraulic transmissivity in m^2/s,
$Q =$ average quantity of fluid discharged per unit time in m^3/s,
$L =$ length of specimen in metres,
$W =$ width of specimen in metres,
$H =$ difference in total head across the specimen in metres.

Ultraviolet resistance

The long-term properties of geotextiles are often threatened by degradation under exposure to sunlight. Since it is practically impossible to install a geotextile without some exposure to sunlight, the UV resistance of a geotextile is critical. The UV resistance of geotextiles is measured in the laboratory by exposing the geotextile specimens to three exposure times under a standard set of humidity and temperature conditions. The exposure times are 150, 300, and 500 hours. One set of specimens is left unexposed as a control. UV radiation is simulated using a xenon arc device. Following the exposure, the specimens are tested using ASTM D 1682 to determine their breaking strength. The results are reported as a percent loss of strength relative to the unexposed specimen.

Geomembranes

Geomembranes are flexible, polymeric sheets that have extremely low permeability and are typically used as liquid or vapour barriers. The types of geomembrane most commonly used for geotechnical applications are the polyethylene (PE) geomembranes. This is due primarily to the high chemical resistance and durability properties of PE geomembranes. High density polyethylene (HDPE) geomembrane is is normally used in important applications and can withstand stresses often imposed on the geomembranes during construction and installation. Very low density polyethylene (VLDPE) is also often used; it has many of the properties of HDPE but is more flexible and can more readily conform to underlying settlements without puncturing. Most PVC geomembranes are generally less expensive than PE geomembranes and can be manufactured in a few large panels to meet the dimensional requirements of specific projects. The large pane sizes allow easier installation since there are fewer numbers to be fabricated in the field.

Sheet properties such as thickness, mass per unit area, tensile behaviour, tear resistance, puncture resistance, environmental stress cracking, ultraviolet resistance, carbon black content, and chemical resistance are the properties that typically need to be considered in the evaluation of the manufactured geomembrane sheet for use in geotechnical applications.

Thickness: For geomembranes with smooth surfaces, the most commonly reported methods are ASTM D 374, ASTM D 751 and ASTM D 1593. Some geomembranes, however, are textured to increase their frictional resistance properties. The texture is either sprayed on to a smooth geomembrane or formed by injecting gases during manufacture of the sheet. Various methods are used for measuring the thickness of a textured geomembrane. For sprayed-on textures the texture may be scraped off and the thickness of the smooth core sheet measured using any of the methods listed. For textured geomembranes formed by gas injection, the minimum thickness may be determined using one of the micrometer methods listed above, or the average thickness may be determined using a mass area method such as ASTM D 1593. Many designers specify that the minimum textured geomembrane thickness be measured using a micrometer since the mass area method may not account for the thin areas between peaks in the texture. However, since the thickness of a textured geomembrane is variable and only limited locations may be measured, even a micrometer may not detect the thinnest portions of the geomembrane. Some designers therefore specify mass area measurement in conjunction with tensile strength testing since tensile strength requirements may not be achieved if the geomembrane is thinner than required.

Mass per unit area: The mass per unit area is determined simply by weighing a measured area of a specimen and then dividing the measured weight by the area measured. This measure is used to characterize the weight of the geomembrane.

Tensile behaviour: The tensile behaviour is measured by performing tests that measure the tensile strength and elongation of the geomembrane. These tests are probably the most commonly performed geomembrane material tests.

The uniaxial tensile tests are performed in a specialized tension testing machine that is capable of accommodating the large elongation-to-break lengths of geomembranes. The test specimen is placed in the machine and gripped with clamps along the top and bottom widths. The specimen is pulled at a constant rate of strain, and the tensile strength and elongation of the specimen are recorded. The results are often plotted as a stress–strain relationship.

The differences among the types of uniaxial tensile tests relate to the standard specimen dimensions, the strain rate, and the material type for which the test is recommended. Multiaxial tests are generally performed when the design application may cause stress on the geomembrane from multiple directions. The test specimen is generally bolted to a circular chamber with pressure exerted on the specimen by air or water. The change in volume of the specimen and the pressure exerted are plotted to create stress–strain curves. Table 3.3 gives typical ranges of tensile strength values of different types of geomembrane.

Table 3.3 *Typical range of selected geomembrane properties*

Property	Tensile strength in machine direction (kN/m)	Puncture resistance (N)	Tear resistance (N)
60-mil HDPE	40-45	300-350	180-200
80-mil HDPE	50-60	400-450	220-270
40-mil VLDPE	20-25	220-290	65-90
60-mil VLDPE	35-40	330-380	110-135
20-mil PVC	8-9	Not available	25-35
40-mil PVC	16-17	Not available	35-55

Tear resistance: During geomembrane installation, the geomembrane may tear due to high winds or handling during panel placement. Tear resistance tests measure a geomembrane's ability to resist tear. The tear resistance determined using ASTM D 1004 is the value most commonly reported by manufacturers for geomembranes.

In the tear test, the test specimen is notched perpendicular to its length. Both ends of the specimen are gripped in a tension machine, the notch being halfway between the grips. As the tensile load is applied, tearing of the specimen begins at the notch and extends across the specimen, approximately perpendicular to the direction of the applied load. The tear resistance is reported as the maximum tensile force attained during the test.

Puncture resistance: Geomembranes are often placed above or below material containing angular or jagged edges. For example in some applications, a granular drainage layer consisting of gravel may be placed above the geomembrane. As load is placed on the granular drainage layer, either by equipment or waste placement, the gravel may be pushed into the geomembrane causing puncture. The benefit of the increased puncture resistance of the geotextile geomembrane systems must be weighed against the potential stability concerns due to the low interface strength between the geomembranes and the geotextiles.

Typically, a geomembrane is considered to have adequate environmental stress crack resistance if it can be subjected to the test conditions for greater than 1000 hours without stress cracking. However, since the time to a forced premature brittle failure has not been correlated to field occurrences of environmental stress cracking, the test is inherently a quality control test related to the manufacture of the geomembrane.

Ultraviolet resistance: Geomembranes that are exposed outdoors may degrade and crack under prolonged exposure. The degradation of outdoor-exposed PE is initiated by thermal and ultraviolet (UV) radiation which lead to the formation of free radicals. The free radicals react rapidly with any oxygen present in the geomembrane. As oxidation degradation progresses, the geomembrane becomes brittle and susceptible to stress cracking. Additives such as carbon black and antioxidants may be added to the PE geomembrane resin to increase the UV resistance of the manufactured sheet. Carbon black absorbs the UV radiation, while the antioxidants act as sacrificial substances to delay the oxidation degradation. The antioxidants have a limited lifetime, however, and will eventually become consumed. A relatively new product that can reduce solar thermal energy absorption and may assist the durability of geomembranes is the white-surfaced geomembrane. PVC geomembranes also contain carbon black to retard the onset of UV degradation; however, progressive brittleness and cracking of PVC geomembranes may also be due to volatilization of the plasticizers. For HDPE geomembranes, typically 2 to 3 percent carbon black content is within acceptable limits while for PVC, approximately 10 to 15 percent carbon black is used. To measure the UV resistance of geomembrane, baseline tension tests are performed on a geomembrane sample. Several samples are also placed under laboratory-simulated UV conditions. At various times within the laboratory simulation, tension tests are performed on some of the test samples. The UV resistance is reported as the percentage of tensile strength retained at a specific time interval.

Chemical resistance: The main purpose of a geomembrane is to act as a barrier to prevent contaminants from penetrating through the geomembrane, and hence the chemical resistance of a geomembrane is extremely important.

Soil–geomembrane interface strength: The interface strength between soils and geomembranes is largely dependent on whether the surface of the geomembrane is flexible or rough enough to push the failure plane into the adjacent soils. Low interface friction values have reported in some case studies of landfill slope failures. For example, the interface friction angle between smooth HDPE geomembranes and clay can be less than 10°. This low interface friction strength can lead to significant stability problems. Also, if the interface between the clay and geomembrane is wetted (i.e., due to condensation of water under the geomembrane, clay swelling, or excess moisture during construction) the interface strength can be further reduced. It is therefore critical that the interface friction tests accurately the model potential field conditions.

Geogrids

Geogrids are high-strength, soil reinforcement products composed of polypropylene, polyethylene, polyester, or PVC – coated polyester. Geogrids are most commonly used to stabilize a slope, construct a soil wall, or to increase the bearing capacity of weak subgrades under embankments. Depending on the requirement of strength in a particular direction, mono-oriented geogrids (for example retaining walls, see Fig. 3.1(f)), bi-directional geogrids (for pavements, increase of bearing capacity etc.) are used (Fig. 3.1(g)). Many of the test methods used for geotextiles and geomembranes also apply to geogrids. The material property is more pertinent to geogrids is the tensile strength of geogrids.

Table 3.4 *Typically range of wide-width geogrid strengths in maximum strength direction*

Geogrid	Tensile strength 5% strain (kN/m)	Ultimate tensile strength (kN/m)
Polyester	130–560	450–1500
Polyester, PVC coated	105–1300	250–400
Polypropylene	10–20	10–35
Polyethylene	25–80	50–140

Tensile strength: The tensile strength values typically reported are from wide-width tensile tests. The wide-width tests are performed for geotextiles with the specimen width typically spanning a few ribs of the geogrid. Specialized jaws are often used to clamp the geogrid properly. Since geogrids are most commonly used to reinforce soil, it is important that the stress–strain properties of geogrids are compatible with the soil. Since soils can begin to fail at approximately 2 to 5 percent of tensile strain in the presence of reinforcement, it is important that most of the geogrid strength is mobilized at low strains. Geogrid strength is therefore often reported in terms of its secant modulus or strength at 2 and 5 percent strains. Table 3.4 presents typical ultimate tensile strengths and tensile strengths at 5 percent strain for various types of geogrids. The allowable tensile strength of a geogrid is typically significantly less than the ultimate tensile strength. The partial factors of safety for installation damage, chemical degradation, and biological degradation range from 1.0 to 1.6, with the partial factor of safety for creep ranging from 1.5 to 3.5.

Geonets and Geocomposite Drainage Nets

Geosynthetic drainage products are extruded sheets of plastic formed into a configuration to promote drainage. They are most commonly composed of PE but may also be composed of polypropylene, polystyrene, high-impact polystyrene, or other materials. Geonets generally consist of two sets of parallel solid or foamed extruded ribs that intersect at a constant angle to form an open net configuration. Channels are formed between the ribs to convey either liquid or gases (Fig. 3.5). Geonets are composed almost exclusively of polyethylene. Since geonets and gecomposite drainage nets are composed of polyethylene and geotextile, many of the properties discussed in earlier apply to these products. However, since the sole purpose of geonets and geocomposite drainage nets is for drainage, the most important properties to be included in specifications are the thickness, crush strength, and transmissivity.

Geosynthetic Clay Liners

Geosynthetic clay liners (GCLs) are very low permeable barriers consisting of a layer of unhydrated, loose granular or powdered bentonite, which is chemically or mechanically adhered to a geotextile or geomembrane (Fig. 3.6). They are generally used as an alternative to compacted clay liners. The differences between GCLs and compacted clay liners are presented in Table 3.5. Due to the inherently

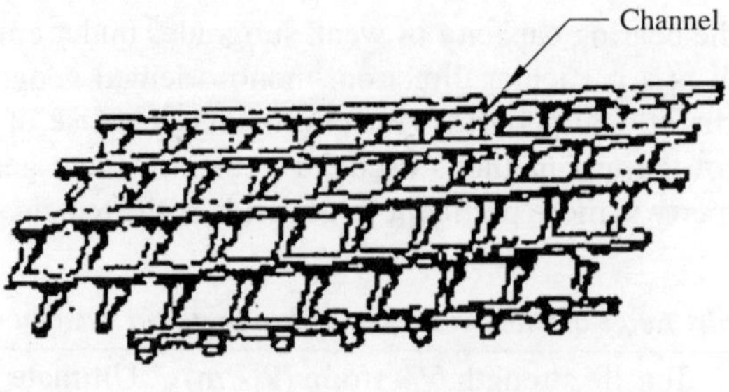

Fig. 3.5 *Geonets*

low shear strength of hydrated bentonite, GCL usage had generally been limited to applications where slope stability of overlying materials has not been of concern.

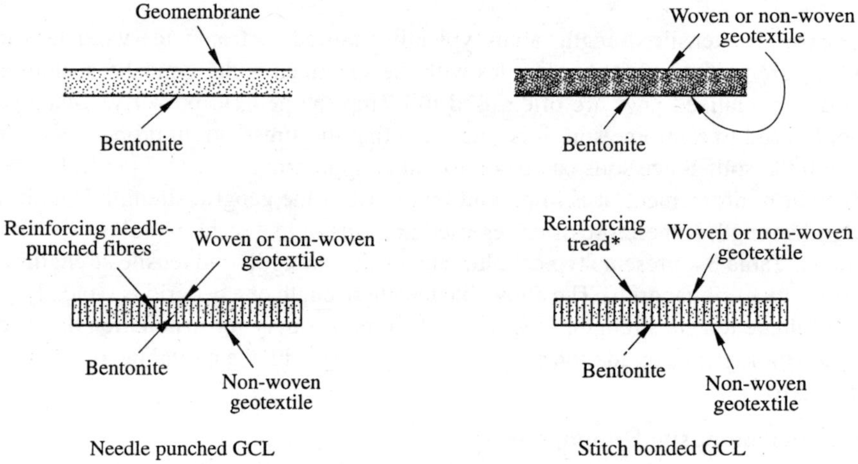

Fig. 3.6 *Geosynthetic clay liners*

Since a GCL is a composite material, typically its properties include the properties of bentonite alone along with the properties of the composite material. The types of geotextile properties important in GCLs include mass per unit area, tensile strength, and puncture resistance.

Bentonite is a clay formed primarily from the mineral montmorillonite. While several types of montmorillonite exist, prominent are calcium and sodium montmorillonite. Montmorillonite crystals are thin, platy and have negative electrical charges associated with their surfaces. Water is strongly attracted to the surface of the negatively charged montmorillonite crystal and is readily absorbed by it. In its unhydrated state, the montmorillonite crystals are densely packed. Once hydrated, the structure becomes very open and swells. The high water absorption and swell characteristics of bentonite lead not only to its low hydraulic conductivity, but also to its low hydrated shear strength.

Free swell test: The free swell test is performed by adding small increments of powdered bentonite to a graduated cylinder over an established period of time. The volume of swell that occurs is measured.

Table 3.5 *Difference between GCLs and compacted clay liners*

Characteristic	Geosynthetic clay liner	Compacted clay liner
Materials	Bentonite clay, adhesives, geotextiles, and geomembranes	Native soils or blend of soil and bentonite
Construction	Manufactured and then installed in the field	Constructed in the field
Thickness	Approximately 10mm	Approximately 0.5 to 1.0m
Hydraulic conductivity of clay	10^{-10} to 10^{-8}cm/s (typical)	10^{-8} to 10^{-7}cm/s (typical)
Speed and ease of construction	Rapid, simple installation	Slow, complicated construction
Water content at time of construction	Essentially dry; cannot desiccate during construction and produces no consolidation water	Nearly saturated; can desiccate and can produce consolidation water
Cost	Less variation	Highly variable depending on site conditions
Experience level	Limited but applications have increased in the recent years	Has been used for many decades

A minimum swell volume of 22 ml is apparently indicative of a bentonite that will achieve a hydraulic conductivity in the 10^{-9} cm/s range; however, there is no definite correlation.

Composite material properties: The types of GCLs currently available can generally be divided into two categories: those with the outer geosynthetic layers mechanically bonded using stitching or needle punching, and those were an adhesive bond is used (Fig. 3.6). The mechanically bonded GCLs are often referred to as reinforced, and their adhesive-bonded counterparts as unreïnforced. The mechanical bonding of reinforced GCLs increases their internal shear strength.

Bentonite content: The bentonite content is simply a measure of the mass of bentonite per unit area of GCL. It is important to note the moisture content of the bentonite when measuring the mass per unit area since it can influence the value. Most GCLs contain between 4 and 5 kg per square metre of bentonite at moisture contents ranging from 12 to 20 percent.

Permeability: Since the main function of a GCL is to act as a low-permeability barrier, its hydraulic conductivity is a very important material property. The hydraulic conductivity of a GCL may be measured using a permeameter. While the hydraulic conductivity varies depending on the effective confining pressure, most GCL manufacturers report a hydraulic conductivity of 5×10^{-9} cm/s.

Tensile strength: As with other geosynthetics, the tensile strength of a GCL may be measured. Both grab tensile and wide-width tensile tests may be performed.

Puncture resistance: Much of the concern regarding the use of GCLs has been that they are prone to puncture due to their thinness. It would obviously be easier to puncture a GCL than a 60 cm thick clay liner. However, GCLs are known to be self-healing. This means that the bentonite will probably swell to fill in the void caused by the puncture.

3.3 Concluding Remarks

In this chapter, a brief review of materials and material properties that are of use in design using soil reinforcement and geosynthetics is presented. Some of the methods have been developed in different countries and specifications have been formulated accordingly. Recognizing this aspect, there have been efforts towards standardization of test procedures in many countries. In USA, the standardization is in advanced stage where as sustained efforts are being made in Europe to develop common European Norms. A partial list of the properties and the relevant ASTM (American Society for Testing of Materials) standards are given in Table 3.6. The behaviour of geosynthetic materials as influenced by temperature is not addressed and the reader is referred to Koerner (1998) and other codes available on the topic.

Table 3.6 *Partial list of ASTM standards*

ASTM Code No.	Code Title
ASTM D4354	Standard practice for sampling of geosynthetics for testing
ASTM D4355	Deterioration of geotextiles from exposure to ultraviolet light and water
ASTM D4439	Standard terminology for geosynthetics
ASTM D4491	Water permeability of geotextiles by permittivity
ASTM D4533	Index trapezoidal tearing strength of geotextiles
ASTM D4545	Standard practice for determining the integrity of factory seams used in joining manufactured flexible sheet geomembranes, peel, shear
ASTM D4594	Effects of temperature on stability of geotextiles
ASTM D4595	Tensile properties of geotextiles by the wide-width strip method
ASTM D4632	Grab breaking load and elongation of geotextiles
ASTM D4716	Determination of the (in-plane) flow rate per unit width and hydraulic transmissivity of a geosynthetic using a constant head
ASTM D4751	Test method for determining apparent opening size of a geotextile
ASTM D4759	Determining the specification conformance of geosynthetics
ASTM D4833	Index puncture resistance of geotextiles, geomembranes and related products
ASTM D4873	Guide for identification, storage and handling of geotextiles
ASTM D4884	Test method for seam strength of sewn geotextiles
ASTM D4885	Test method for determining performance tensile strength of geomembranes using wide-strip testing
ASTM D4886	Abrasion resistance of geotextiles (sandpaper/sliding block method)
ASTM D5035	Breaking strength and elongation of textile fabrics (strip method)
ASTM D5101	Measuring the soil geotextile system clogging potential by the gradient ratio

Continued on next page

Table 3.6 *Continued*

ASTM Code No.	Code Title
ASTM D5141	Determination of filtering efficiency and flow rate of a geotextile for silt fence application using site specific soil
ASTM D5199	Measurement of nominal thickness of geotextiles and geomembranes
ASTM D5261	Measuring mass per unit area of geotextiles
ASTM D5262	Evaluation of unconfined tension creep behaviour of geosynthetics
ASTM D5321	Determination of the coefficient of soil and geosynthetic or geosynthetic and geosynthetic friction by the direct shear method
ASTM D5322	Immersion procedures for evaluating the chemical resistance of geosynthetics to liquids
ASTM D5323	Determination of 2% secant modulus for polyethylene geomembranes
ASTM D5493	Test method for the permittivity of geotextiles under load
ASTM D5494	Test methods for the determination of pyramidal puncture resistance of unprotected and protected geomembranes
ASTM D5496	Practice for in situ immersion testing of geosynthetics
ASTM D5514	Large scale hydrostatic puncture testing of geosynthetics
ASTM D5567	Hydraulic conductivity ratio (HCR) testing of soil/geotextile systems
ASTM D5617	Multi-axial tension test for geosynthetics
ASTM D5721	Air oven aging of polyolefin geomembranes
ASTM D5747	Evaluation of chemical resistance of geomembranes to liquids
ASTM D5818	Obtaining samples of geosynthetics from a test section for assessment of installation damage
ASTM D5819	Selecting test methods for experimental evaluation of geosynthetic durability
ASTM D5820	Pressurized air channel evaluation of dual seamed geomembranes
ASTM D5884	Tearing strength of internally reinforced geomembranes
ASTM D5885	Oxidative induction time of polyolefin geosynthetics by high pressure differential scanning calorimetry
ASTM D5887	Measurement of index flux through saturated geosynthetic clay liner specimens using a flexible wall permeameter
ASTM D5888	Storage and handling of geosynthetic clay liners
ASTM D5889	Quality control of geosynthetic clay liners
ASTM D5890	Swell index of clay mineral component of geosynthetic clay liners
ASTM D5891	Fluid loss of clay component of geosynthetic clay liners
ASTM D5970	Deterioration of geotextiles from outdoor exposure
ASTM D5993	Measuring the mass per unit area of GCL
ASTM D5994	Measuring the core thickness of textured geomembranes

Continued on next page

Table 3.6 *Continued*

ASTM Code No.	Code Title
ASTM D6072	Obtaining samples of geosynthetic clay liners
ASTM D6088	Installation of geocomposite pavement drains
ASTM D6102	Installation of geosynthetic clay liners
ASTM D6213	Evaluation of chemical resistance of geogrids to liquids
ASTM D6214	Determination of the integrity of field seams used in joining geomembranes by chemical fusion methods
ASTM D6241	Static puncture strength of geotextiles using a 50-mm probe
ASTM D6243	Determination of internal and interface shear resistance of GCL by the direct shear method
ASTM D6364	Determination of short-term compression behaviour of geosynthetics
ASTM D6388	Evaluation of chemical resistance of geonets to liquids
ASTM D6389	Evaluation of chemical resistance of geotextiles to liquids
ASTM D6392	Integrity of nonreinforced geomembrane seams produced using thermo-fusion methods.
ASTM D6434	Standard guide for the selection of test methods for flexible polypropylene (fPP) geomembranes
ASTM D6454	Standard test method for determining the short-term compression behaviour of turf reinforcement mats
ASTM D6495	Acceptance testing requirements for geosynthetic clay liners
ASTM D6496	Determining average bonding peel strength between the top and bottom layers of needle-punched geosynthetic clay liners
ASTM D6497	Mechanical attachment of geomembrane to penetrations or structures
ASTM D6523	Evaluation and selection of alternative daily covers for sanitary landfills
ASTM D6524	Standard test method for measuring the resiliency of turf reinforcement mats
ASTM D6525	Standard test method for measuring nominal thickness of permanent rolled erosion control products
ASTM D6566	Test method for measuring mass per unit area of turf reinforcement mats
ASTM D6567	Test method for measuring the light penetration of a turf reinforcement mat (TRM)
ASTM D6574	Test method for determining the (in-plane) hydraulic transmissivity of a geosynthetic by radial flow.
ASTM D6575	Test method for determining stiffness of geosynthetics used as turf reinforcement mats (TRM's)

Bibliography

[1] BS 8006 (1995) British Standard: Code of practice for strengthened/reinforced soils and other fills.

[2] Koerner, R.H. (1998) *Designing with Geosynthetics*, 4th Edition, Printice Hall, Englewood Cliffs, N.J.

[3] Rao, G.V. (1990) *Engineering with Geosynthetics* (Rao, G. Venkatappa, G.V.S. Suryanarayana Raju), Tata McGraw-Hill Publishing Company Limited, New Delhi.

[4] Rao, G.V and K. Balan (2000) *Coir Geotextiles, Emerging Trends*, The Kerala State Coir Corporation Ltd., Alappuzha, India.

[5] Rao, P.J. (1996) *Proceedings of the National Seminar on Jute Based Geotextiles.*

[6] Sharma, H.D. and S.P. Lewis (1994) *Waste Containment systems, Waste Stabilization, and Landfills: Design and Evaluation*, John Wiley & Sons Inc., USA.

4 Improvement of Bearing Capacity

4.1 Introduction

One of the areas where reinforced soil technique could be used very effectively is in the improvement of bearing capacity of foundation soils. Studies by several investigators (for example, Binquet and Lee, 1975a, 1975b, Guido et al. 1986) clearly indicate the advantages and possibilities of improving the bearing capacity and stiffness (in terms of the load–settlement behaviour) by reinforcing the foundation soil in the case of sand. Considerable research work using experimental, analytical and numerical methods is available in literature. A number of experimental and theoretical studies have been carried out for evolving methodologies for determining the improvement of bearing capacity in terms of bearing capacity ratio (BCR). BCR is defined as the ratio of improved (or required) bearing capacity to the bearing capacity of foundation soil without reinforcement. A number of investigators (Binquet and Lee, 1975, Guido et al. 1986, Singh, 1988, Bindumadhava, 1990, Sridharan 1990, Srinivasa Murthy et al. 1993, Shivashankar et al. 1993, Kumar and Saran, 2001) developed design methods using different forms of reinforcement, types of footing and soil conditions. A modification of Binquet and Lee's approach for strip footings was made by Singh (1988) and Srinivasa Murthy et al. (1993), which extended the anlysis to rectangular, square and circular footings. The tension mobilized in the reinforcement is taken as a function of the mobilized interfacial frictional angle and the total normal force on the plan area of the reinforcement beyond the assumed slip surface. A method was also suggested to determine the mobilized friction as a function of strain levels. Of the above methods, the foremost method is due to Binquet and Lee (1975) and is explained in the following sections. When some form of reinforcement is used in the foundation soil, it is necessary to identify the modes of failure due to the presence of reinforcement so that the effectiveness of the reinforced foundation is determined. To develop a design procedure, Binquet and Lee assumed that the following mechanisms of failure are likely to occur in a reinforced soil foundation.

4.2 Modes of Failure

Bearing capacity of shallow foundations resting on reinforced soil has been studied in detail by Binquet and Lee (1975a, b). According to them, if layers of reinforcing strips or ties are placed in the soil under a shallow strip foundation, the nature of failure in the soil mass is similar to the pattern shown in Fig. 4.1(a). Such a failure generally occurs when the first layer of reinforcement is placed at a depth d greater than about $2/3B$ (B = width of the foundation). If the reinforcements in the first layer are strong and are sufficiently concentrated, they may act as a rigid base at a limited depth. The type of failure shown in Fig. 4.1(b) could occur if d/B is less than about 2/3 and the number of layers of reinforcement (N) is less than about 2 to 3. In this type of failure, reinforcement-tie pullout occurs. The most beneficial effect of reinforced earth is obtained when d/B is less than about 2/3, and the number of reinforcement layers is greater than 4 but not more than 6 to 7. (The ties must be sufficiently long.) In this case the soil mass fails when the upper ties break Fig. 4.1(c).

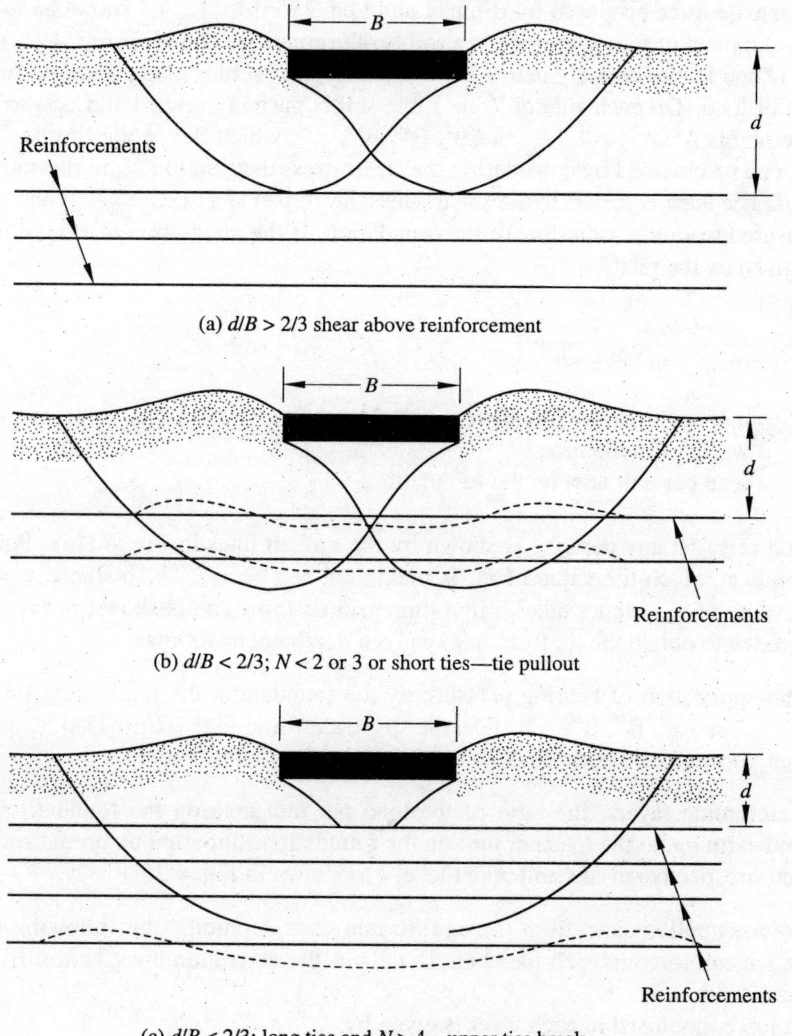

Fig. 4.1 *Three modes of bearing capacity failure in reinforced earth (redrawn after Binquet and Lee, 1975)*

4.3 Determination of Force Induced in Reinforcement Ties

4.3.1 Location of failure surface

When designing shallow strip foundations, an engineer must estimate the force that develops in the reinforcing ties as a result of the foundation load. This section presents the analytical procedure for estimating the tensile force proposed by Binquet and Lee (1975b). Fig. 4.2 shows an idealized condition for the development of failure surface in a soil for the condition shown in Fig. 4.1(c). It consists of a central zone (Zone I) immediately below the foundation that settles along with the foundation with the application of load. On each side of Zone I, the soil is pushed outward and upward and this area is Zone II. The points A′, A″, A‴,..., and B′, B″, B‴, ..., which define the limiting lines between Zones I and II, can be obtained by considering the shear stress distribution τ_{xz} in the soil caused by the foundation load. The term τ_{xz} refers to the shear stress developed at a depth z below the foundation at a distance x measured from the centerline of the foundation. If the integration of Boussinesq's equation is used, τ_{xz} is given by the relation

$$\tau_{xz} = \frac{4xq_R xz^2}{\pi\left[(x^2+z^2-b^2)^2+4b^2z^2\right]} \tag{4.1}$$

where b = half-width of the foundation = $B/2$,
B = width of foundation,
q_R = load per unit area on the foundation.

The variation of τ_{xz} at any depth z, is shown by the broken lines in Fig. 4.2(a). Points A′ and B′ refer to the points at which the value of τ_{xz} is maximum at $z = z_2$. The distance $x = X_0$ at which the maximum value of τ_{xz} occurs takes a non-dimensional form and is shown in Fig. 4.2(b). Other assumptions needed to obtain the tie force at any given depth are as follows:

1. Under the application of bearing pressure by the foundation, the reinforcing ties at points A′, A″, A‴, ..., and B′, B″, B‴, ..., take the shape shown in Fig. 4.2(c). That is, the ties take two right angle turns on each side of Zone I around two frictionless rollers.

2. For N reinforcing layers, the ratio of the load per unit area on the foundation supported by reinforced earth q_R, to the load per unit on the foundation supported by unreinforced earth q_0, is a constant irrespective of the settlement level s as shown in Fig. 4.2(d).

With the above assumptions and from the equilibrium considerations, the following relationship is derived for the tension force in each member. Details of the derivation are given in Binquet and Lee (1975) and Das (1999).

The tension force mobilised in each layer is given by

$$T_{(N)} = \left(\frac{q_0}{N}\right)\left(\frac{q_R}{q_0}-1\right)(A_1B - A_2\Delta H), \tag{4.2}$$

where q_0 is the bearing capacity of unreinforced soil and A_1 and A_2 are the functions of z obtained from Fig. 4.3.

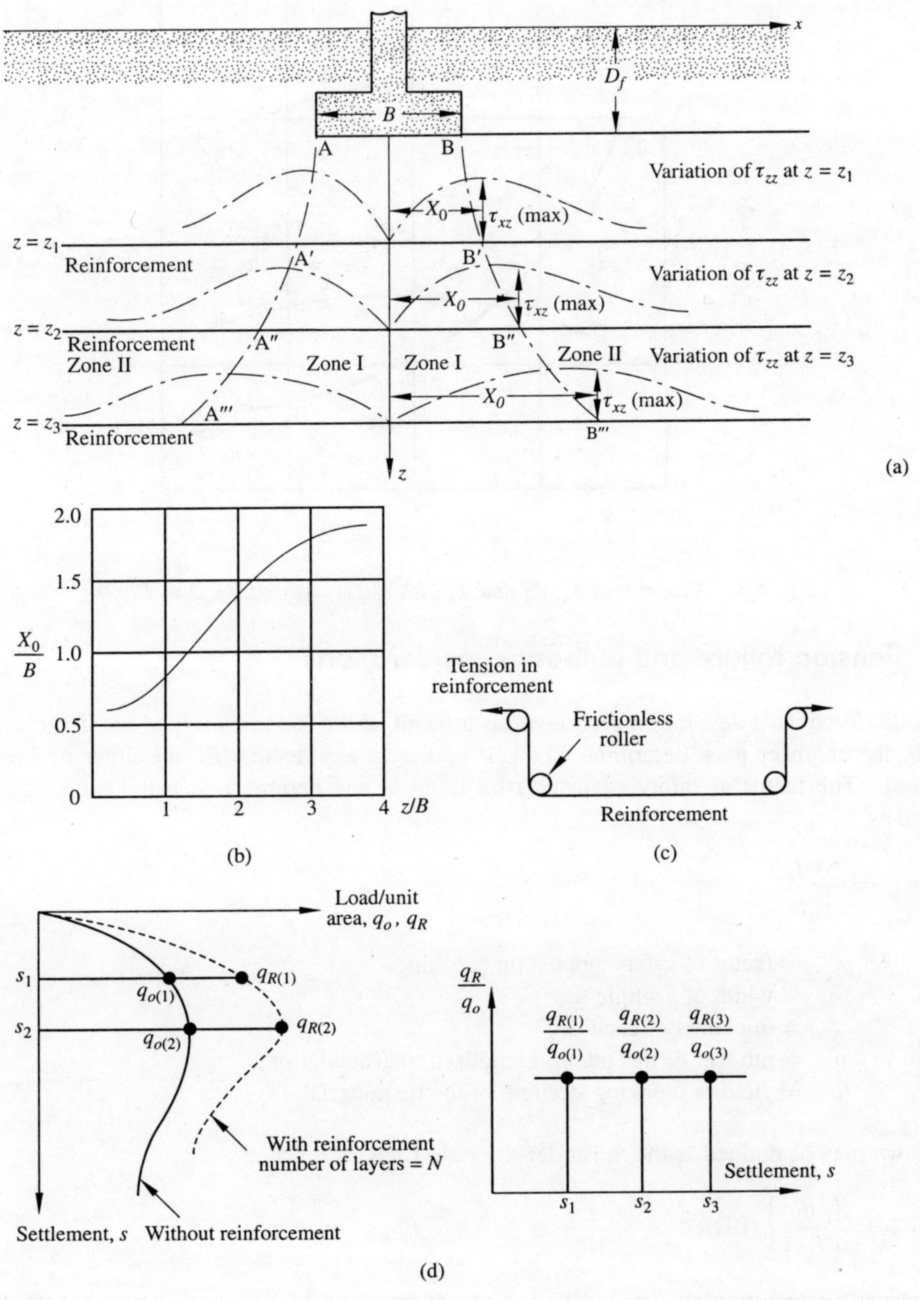

Fig. 4.2 *(a) Variation of shear stress with depth; (b) variation of X_0/B with z/B; (c) tension in the reinforcement in reinforced soil under footing; (d) relationship between load per unit area and settlement for foundations resting on unreinforced and reinforced soil (Das, 1999).*

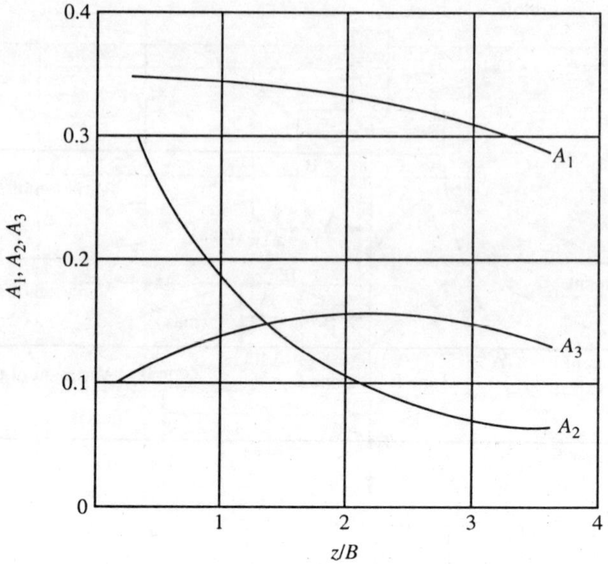

Fig. 4.3 *Variation of A_1, A_2 and A_3 with z/B (redrawn from Das, 1999)*

4.3.2 Tension failure and pull-out considerations

Once the tie forces that develop in each layer as a result of the foundation load are determined from Eq. (4.2), the engineer must determine whether the ties at any depth will fail either by breaking or by pull-out. The factor of safety against tie-breaking at any depth z below the foundation can be calculated as

$$FS_{(B)} = \frac{wtnf_y}{T_{(N)}} \tag{4.3}$$

where $FS_{(B)}$ = factor of safety against tie-breaking,
w = width of a single tie,
t = thickness of each tie,
n = number of ties per unit length of the foundation,
f_y = yield or breaking strength of the tie material.

The term wn may be defined as the linear density ratio LDR, and hence

$$FS_{(B)} = \left[\frac{tf_y}{T_{(N)}}\right] (\text{LDR}). \tag{4.4}$$

The resistance against the tie being pulled out comes from the frictional resistance between the soil and the ties at any depth and is given by

$$F_B = 2 \tan \phi_\mu \quad [\text{normal force}]$$

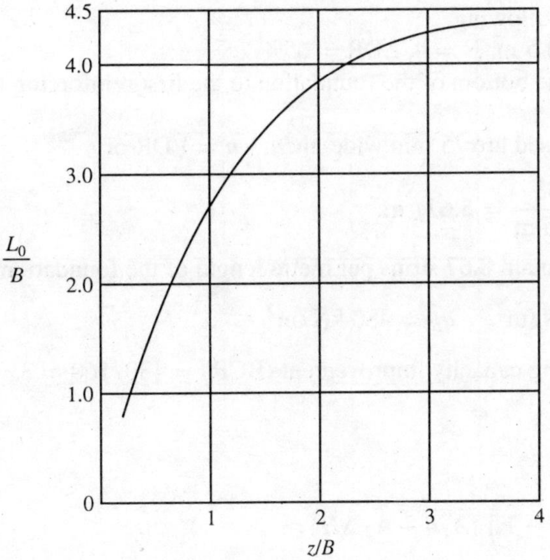

Fig. 4.4 *Variation of L_0/B with z/B (Redrawn from Das, 1999)*

$$F_B = \tan\phi_\mu \left[(\text{LDR}) \int_X^{L_0} \sigma(q_R)dx + \text{LDR}(\gamma)(L_0 - (X_0))(z + D_f) \right], \tag{4.5}$$

where γ is the unit weight of soil, D_f the depth of foundation and ϕ_μ the tie–soil friction angle.

The value of $x = L_0$ is generally assumed to be the distance at which $\sigma(q_R)$ equals $0.1q_R$. The value of L_0 as a function of depth z is given in Fig. 4.4. Equation (4.5) may be simplified as

$$F_B = 2\tan\phi_\mu \left[(\text{LDR}) \left[A_3 B q_0 \left(\frac{q_R}{q_0} \right) + (\gamma)(L_0 - X_0)(z + D_f) \right] \right], \tag{4.6}$$

where A_3 is a non-dimensional quantity that may be expressed as a function of depth (z/B) (see Fig. 4.3). The factor of safety against tie pull-out, $\text{FS}_{(P)}$, is

$$\text{FS}_{(P)} = F_B/T_N. \tag{4.7}$$

The approach is illustrated by a simple example.

Example 4.1 Design a continuous foundation that will carry a load of 480 kN/m. Use the following parameters.

Soil: $\gamma =$ kN/m^3; $\phi = 30°$; $E_s = 3 \times 10^4$ kN/m^2; $\mu_s = 0.35$

Reinforcement ties: $f_y = 2.5 \times 10^5$ kN/m^2; $\phi_\mu = 20°$; $\text{FS}_{(B)} = 3$; $\text{FS}_{(P)} = 2.0$

Foundation: $D_f = 1$ m; allowable bearing capacity of the foundation = 160 kN/m^2; design life of structure = 50 years.

Solution Assume the following:

B = 1 m, $\Delta H = 0.5$ m, $N = 4$, LDR = 65%

d = depth from the bottom of the foundation to the first reinforcing layer = 0.5 m.

If the reinforcing strips used are 75 mm wide, then, wn = LDR or

$$n = \frac{\text{LDR}}{w} = \frac{0.65}{0.075 \text{ m}} = 8.67/\text{ m}.$$

Hence each layer will contain 8.67 strips per metre length of the foundation.

$$q_0 = q_{\text{all}} = 160 \text{ kN/m}^2, \quad q_R = 480 \text{ kN/m}^2$$

Hence the required bearing capacity improvement (BCR) = 480/160 = 3.

Calculation of tie force

From Eq. (4.2),

$$T_{(N)} = \left(\frac{q_0}{N}\right)\left(\frac{q_R}{q_0} - 1\right)(A_1 B - A_2 \Delta H).$$

The tie force for each layer, calculated using the above equation is given in Table 4.1.

Table 4.1 *Calculation of tensile forces in reinforcement members*

Layer No.	$\left(\frac{q_0}{N}\right)\left(\frac{q_R}{q_0} - 1\right)$	z(m)	$\frac{z}{B}$	$A_1 B$	$A_2 \Delta H$	$A_1 B - A_2 \Delta H$	$T_{(N)}$ (kN/m)
1	80	0.5	0.5	0.35	0.125	0.225	18.00
2	80	1.0	1.0	0.34	0.090	0.250	20.00
3	80	1.5	1.5	0.33	0.065	0.265	21.20
4	80	2.0	2.0	0.32	0.050	0.270	21.60

Note: A_1 and A_2 are from Fig. 4.3. $B = 1$ m; $\Delta H = 0.5$ m; $q_R/q_0 = 480/160 = 3$

Calculation of tie resistance due to friction, F_B

Use Eq. (4.6).

$$F_B = 2\tan\phi_\mu \left[(\text{LDR})\left[A_3 B_{q_0}\left(\frac{q_R}{q_0}\right) + (\gamma)(L_0 - X_0)(z + D_f)\right]\right]$$

The values calculated using the above equation are given in Table 4.2. It may be noted that the factor of safety against pull-out of the tensile reinforcement varies from 2.21 to 4.15 depending on the location.

Calculation of tie thickness to resist tie-breaking

$$\text{FS}_{(B)} = \frac{tf_y}{T_{(N)}}(\text{LDR}).$$

Here $f_y = 2.5 \times 10^5$ kN/m², LDR = 0.65, and $\text{FS}_{(B)} = 3$. So,

$$t = \frac{\text{FS}_{(B)} T_{(N)}}{(\text{LDR})(f_y)}.$$

Table 4.2 *Calculation of pull-out lengths*

	Layer number			
Quantity	1	2	3	4
$2\tan\phi_\mu$(LDR)	0.473	0.473	0.473	0.473
A_3	0.125	0.14	0.15	0.15
$A_3 B q_0 (q_R/q_0)$	60.0	67.2	72.0	72.0
z (m)	0.5	1.0	1.5	2.0
z/B	0.5	1.0	1.5	2.0
L_0 (m)	1.55	2.6	3.4	3.85
X_0 (m)	0.55	0.8	1.1	1.4
$L_0 - X_0$ (m)	1.0	0.8	1.1	1.4
$z + D_f$ (m)	1.5	2.0	2.5	3.0
$\gamma(L_0 - X_0)(z + D_f)$	24.0	57.6	92.0	117.6
F_B (kN/m)	39.73	59.03	77.6	89.68
$FS_{(P)} = F_B/T_{(N)}$	2.21	2.95	3.66	4.15

Note: A_3 is from Fig. 4.3; X_0 from Fig. 4.2; L_0 is from Fig. 4.4; $T_{(N)}$ is from the preceding table.

So, for layer 1

$$t = \left[\frac{3}{0.65 \times (2.5 \times 10^5)}\right] T_{(N)} = (1.846 \times 10^{-5}) T_{(N)}$$
$$= (1.846 \times 10^{-5})(18) \times 1000 = 0.33 \text{ mm.}$$

So, for layer 2,

$$t = (1.846 \times 10^{-5})(20) \times 1000 = 0.37 \text{ mm.}$$

Similarly, for layer 3, $t = 0.39$ mm and for layer 4, $t = 0.40$ mm.

If galvanized steel is used, the rate of corrosion is about 0.025 mm/year. So t should be $0.40 + (0.025)(50) = 1.65$ mm. Normally, a minimum thickness of 6 mm is available commercially. Hence strips 75 mm wide and 6 mm long are considered as reinforcement for the present case.

4.3.3 Calculation of minimum length of ties

The minimum length of ties in each layer should equal $2L_0$. The design could be changed by varying B, d, N, and ΔH to determine the most economical combination. It can be observed that the pull-out lengths are considerably high. In the above example, the maximum length for the fourth layer of reinforcement is 7.7 m, which is rather long. This is due to the fact that the pull-out resistance is available from the interfacial friction between the soil and reinforcement only. In order to increase the pull-out resistance and reduce the pull-out lengths, grid reinforcement in the form of geosynthetic grids or steel grids and anchored reinforcements have been developed and used successfully.

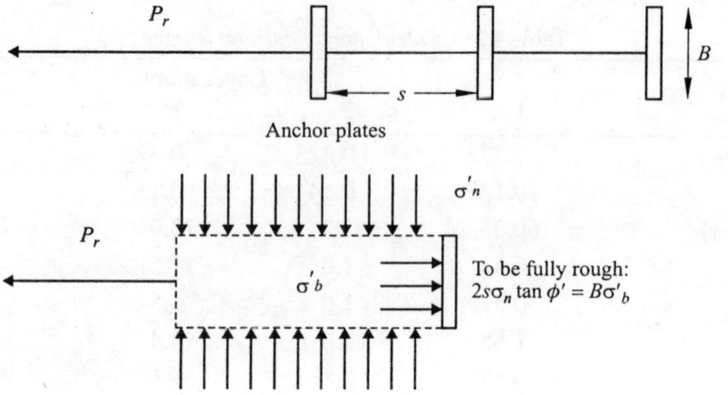

Fig. 4.5 *Mechanism of pull-out resistance due to anchors or bearings (Jewell, 1996)*

Srinivasa Murthy et al. (1993) and Kumar and Saran (2001) point out that there are limitations in the assumptions of the Binquet and Lee theory with regard to variation of soil-reinforcement friction angle. Soil-reinforcement friction angle was assumed as constant for all layers, whereas detailed analysis showed that this angle is a function of strain level at different depths. Elastic theory is used for simplicity, whereas use of nonlinear theories would have given a better picture. In spite of the limitations of these assumptions, the theory provides a basis for improving the bearing capacity of sands using soil reinforcement.

4.3.4 Curtailment of pull-out length of reinforcement

In the above example it can be observed that the pull-out length is appreciable, and that the total length of reinforcement in the bottom layers is about 7.7 m. This can be reduced by providing anchors or strips in the transverse direction at suitable locations so that the pull-out or bond resistance is improved, and at the same time the total length of the reinforcement is reduced.

Assume that the two strips are welded on either side of the main reinforcement. These strips or anchor plates provide bearing resistance as indicated in Fig. 4.5. Consider two anchor plates placed at a distance S; the pull-out resistance is mobilised over a strip of length S with thickness B per unit length. If the surface of the strip is rough, the mobilised bond resistance is given by (Jewell, 1996)

$$B\sigma'_b = 2S\sigma'_n \tan\phi', \tag{4.8}$$

where σ'_b is the bond stress, σ'_n is the effective normal stress and ϕ' is the friction angle of the backfill. If strips of thickness t are used in both the directions, B equals $3t$.

$$\frac{\sigma'_b}{\sigma_n} = \frac{2S}{B}\tan\phi'. \tag{4.9}$$

A theoretical relationship between bond stress ratio σ'_b/σ'_n and the backfill soil properties obtained from limit equilibrium considerations using bearing capacity theory (Jewell, 1996) is given by,

$$\frac{\sigma'_b}{\sigma'_n} = \tan(\pi/4 + \phi'/2)e^{(\pi/2+\phi')\tan\phi'}. \tag{4.10}$$

For $\phi' = 30°$,

$$\frac{\sigma'_b}{\sigma'_n} = \tan(45 + 15)e^{1.209} = 5.8.$$

This is the ultimate bond stress ratio. One can use a factor of safety of 1.50 to obtain the safe bond stress ratio and equate it to $2S/B \tan\phi'$ (Eq. (4.9)) to obtain a suitable combination of spacing and width of anchor plate. Assuming that B is 100 mm, the spacing S can be obtained from

$$\text{Safe } \frac{\sigma'_b}{\sigma'_n} = \frac{5.8}{1.5} = 3.867 = \frac{2S}{100} \tan\phi' \Rightarrow S = 334.86 \text{ m.}$$

Hence $S = 300$ mm is taken as appropriate. The layout of the reinforcement with anchor plates is shown in Fig. 4.6. The length of the reinforcement is taken as 3 m (1.5 m on either side of the centre-line) and 10 anchor plates on the reinforcement, five on either side at a spacing of 300 m C/C (centre-to-centre distance) are provided, as shown in Fig. 4.6. To examine if the above combination of total length of reinforcement, spacing and width of anchor plates is satisfactory, calculations were performed the results of which are given in Table 4.3.

Table 4.3 *Calculation of bond resistance of anchor plates*

Depth from ground surface $(z + D_f)$, m	σ_n(kN/m^2) (kN/m^2)	Bond resistance per anchor $\sigma_b = \sigma_n \times 3.867$	Tensile force per metre length from Table 4.2 (kN/m)	Factor of safety against pull-out
1.5	24	92.8	18	5.16
2.0	32	123.7	20	6.19
2.5	40	154.7	22	7.03
3.0	48	185.6	22.4	8.29

Table 4.3 shows that the factors of safety against pull-out are higher and hence satisfactory. The above example is only an illustration; many combinations of dimensions can be worked out so that an optimised layout is obtained. Advantages of providing pull-out resistance as illustrated above are realised and different types of geogrids are available nowadays for use.

4.4 Guidelines on the Use of Geogrids

Based on a comprehensive review of experimental data, analytical studies and economical considerations, Wayne et al. (1998) suggested guidelines for the dimensions and locations of geogrids for improving the bearing capacity in good frictional soils. These are useful for 0.6 to 1.0 m wide strip footing and 1.5 to 1.8 m square footings for a bearing capacity ratio in the range of 1.5 to 2.5, and valid for good frictional material as backfill. A minimum of 10 cm of sand cushion between the bottom-most layer and the native soil is necessary to generate adequate interfacial frictional resistance. Some guidelines are given in Table 4.4. It is necessary to examine the applicability of the above guidelines for each specific case in improvement of bearing capacity by proper material testing and analysis.

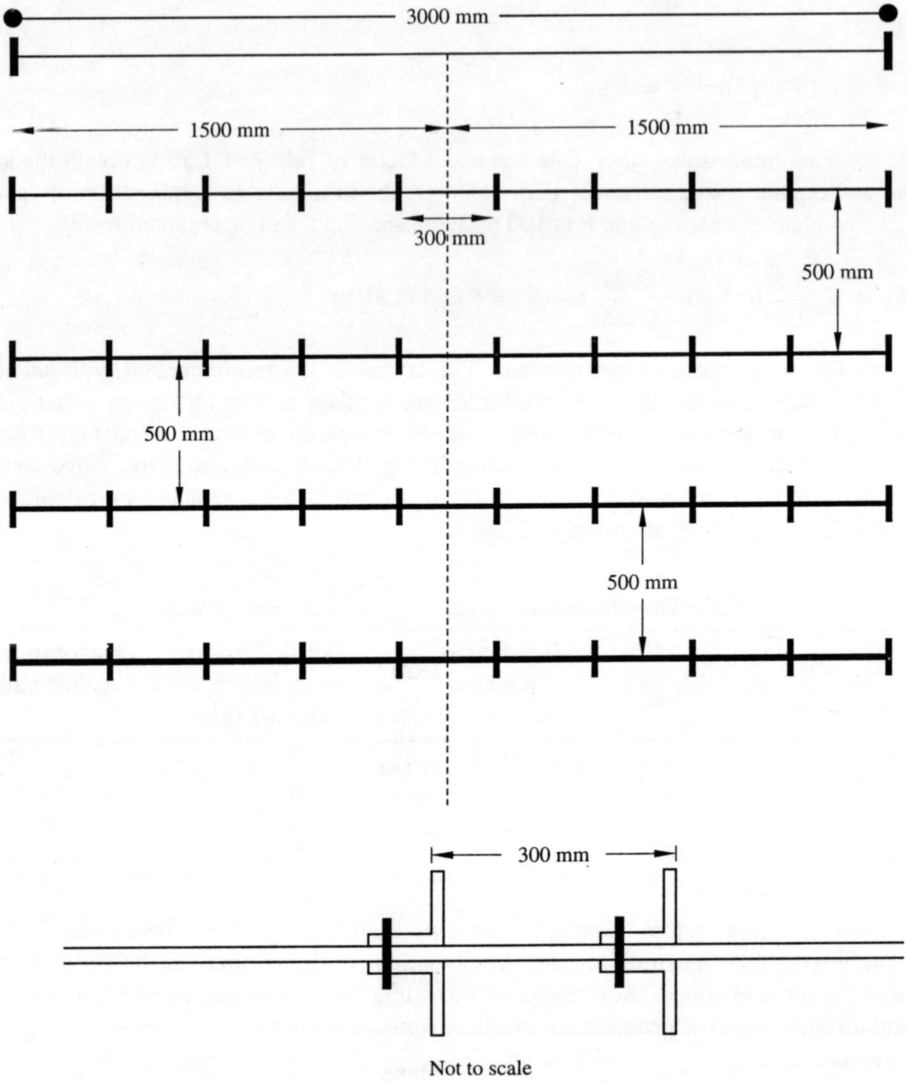

Fig. 4.6 *Schematic diagram of reinforcement with anchor plates in soil below the foundation*

4.5 Bearing Capacity Improvement in Soft Soils

Reinforcement in soft soils for improving bearing capacity and reducing settlement is being increasingly done with a combination of vertical inclusions such as stone columns and horizontal layers of geosynthtics. This combined system is often referred to as Geosynthetic Reinforced Foundation Bed (GRFB) and is used in many areas of soft soils in Japan.

Table 4.4 *Guidelines for use of bi-axial geogrids for bearing capacity improvement (Wayne et al., 1998)*

	Typical value	Recommended (not greater than)
u	$0.15B$ to $0.3B$	$0.5B$
s	$0.15B$ to $0.3B$	$0.5B$
z	$0.5B$ to $1.0B$	$2.0B$
b	$2.0B$ to $3.0B$	$4.0B$
a	$0.1B$ to $0.2B$	$0.3B$
Δl	$0.5B$ to $1.0B$	$2.0B$
N	2 to 4	5

Note: B–footing width; u–distance from footing base to uppermost geogrid; s–spacing between geogrid layers; z–thickness of reinforced fill; b–width of reinforced fill; a–distance from lowest geogrid to bottom of reinforced fill; Δl–length of geogrid extended beyond each end of the strip footing; N–number of geogrid layers

Shivashankar et al. (1993) studied the improvement in bearing capacity of footings resting on granular bed overlying soft clay, assuming a punching shear failure mechanism in the foundation soil. The improvement is attributed to three effects: (a) shear layer effect, (b) confinement effects due to the interaction between sand and reinforcement in the sand layer, and (c) additional surcharge effects. The approach is simple and elegant and is discussed in the following sections.

The ultimate bearing capacity of a footing resting on soft soil is given by

$$q_u = c_u N_c. \tag{4.11}$$

When a granular bed of thickness H of bulk density γ_s and friction angle ϕ_s with reinforcement is provided over soft soil, the bearing capacity of the footing resting on this foundation medium is increased. Frictional forces developed between the soil and the reinforcement induce tensile strains in the reinforcement. The tensile strains developed provide the confining effect. This will induce additional shearing resistance along the vertical plane at the edge, which exponentially decrease with distance away from the edge of the footing. The three effects contribute to increase in bearing capacity, given as

$$q_u + \Delta q_R = c_u N_c + \Delta q_{SL} + \Delta q_{CE} + \Delta q_{SE}. \tag{4.12}$$

The terms, $\Delta q_R \Delta q_{SL} + \Delta q_{CE} + \Delta q_{SE}$ denote the improvements in bearing capacity due to reinforcement, shear layer effect, confining layer effect and surcharge effect, respectively. The improvement in bearing capacity, defined in terms of bearing capacity ratio BCR as the ratio of improved bearing capacity to the original bearing capacity is given as

$$\text{BCR} = 1 + \Delta\text{BCR}_{SL} + \Delta\text{BCR}_{CE} + \Delta\text{BCR}_{SE}, \tag{4.13}$$

where $\Delta\text{BCR}_{SL} + \Delta\text{BCR}_{CE} + \Delta\text{BCR}_{SE}$ are the increases in bearing capacity due to shear layer effect, confinement effect and surcharge effect. The improvements are worked out for each effect considering the mechanics indicated in Fig. 4.7.

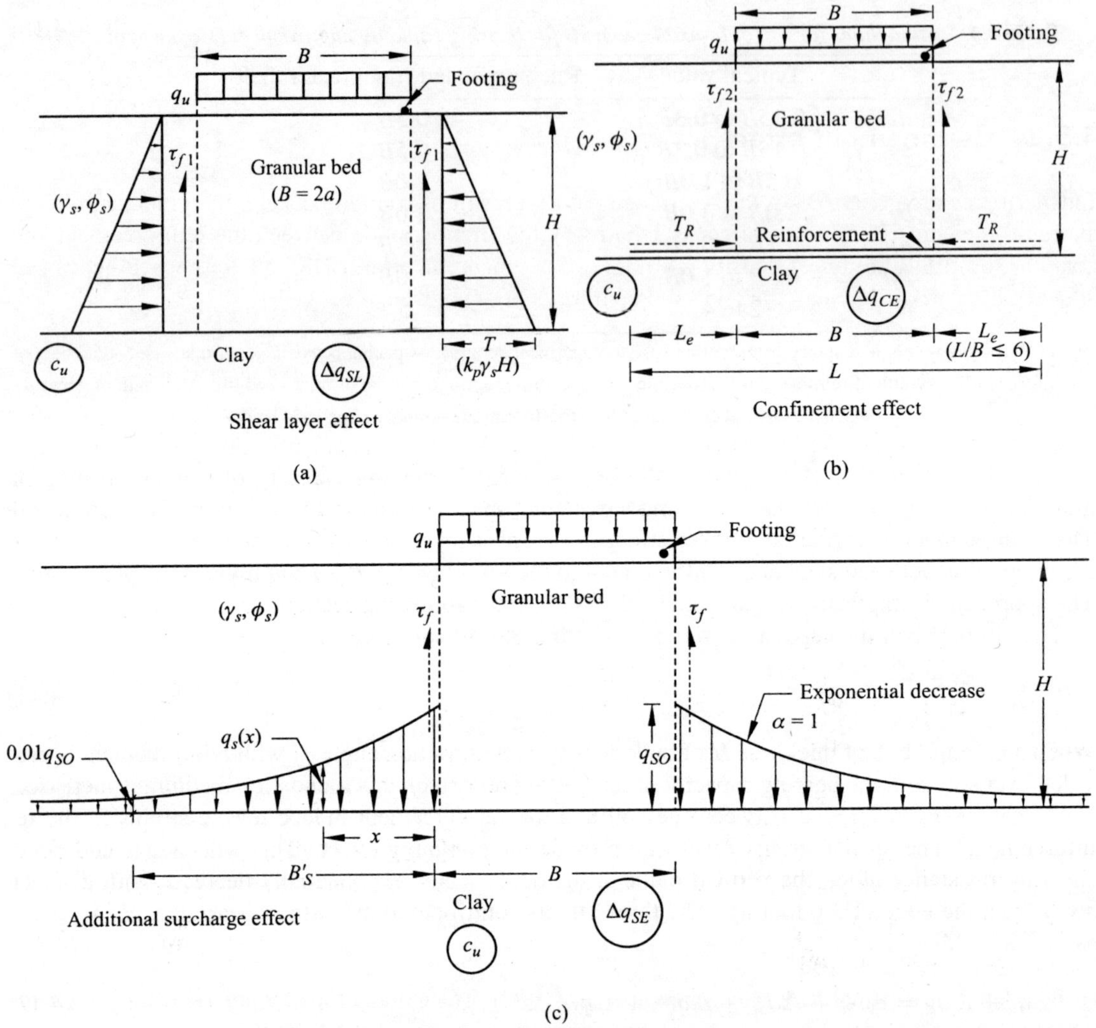

Fig. 4.7 *Improvement of bearing capacity due to (a) shear layer effect (b) confinement effect (c) surcharge effect (Shivashankar et al., 1993)*

4.5.1 Shear layer effect

The shearing stresses that are developed along the vertical plane at the edge of the footing are given by Fig. 4.7(a).

$$\tau_{f1} = \frac{k_p \gamma_s H^2}{2} \tan \phi_s \tag{4.14}$$

$$\Delta q_{SL} = \frac{2\tau_{f1}}{B}$$

$$\Delta \text{BCR}_{SL} = \frac{2\tau_{f1}}{BN_c c_u} \tag{4.15}$$

4.5.2 Confinement effect

Tensile forces are generated in the reinforcement as a result of friction between the granular soil and the reinforcement. If L_e is the effective length, ϕ_R the friction angle between the reinforcement and granular soil, LDR the linear density ratio of reinforcement material (LDR = 1 for geosynthetics and 0.5 to 0.7 for metallic grids), the reinforcement force (T_R) is given as

$$T_R = (\gamma_s H) \tan \phi_R L_e (\text{LDR}). \tag{4.16}$$

L_e can be taken as equal to $2.5B$ or greater. The frictional resistance (T_{f2}) developed at the edge of the footing due to the effect of reinforcement (Fig. 4.7 *b*) leading to confinement is given as

$$\tau_{f2} = T_R \tan \phi_B.$$

Hence

$$\Delta q_{CE} = \frac{\tau_{f2}}{B}, \quad \text{and}$$

$$\Delta \text{BCR}_{CE} = \frac{2\tau_{f2}}{Bc_u N_c} \tag{4.17}$$

4.5.3 Surcharge effect

Shivashankar et al. (1993) showed that for an exponentially decreasing surcharge on either side from the edge of the footing (Fig. 4.7 c), the surcharge stress (q_{SO}) at the edge of the footing is given as

$$q_{SO} = 0.84(\Delta q_{SL} + \Delta q_{cE}). \tag{4.18}$$

Further details are available in Shivashankar et al. (1993). The authors found good agreement between the experimental observations and the predictions made using the above formulation.

4.6 Concluding Remarks

This chapter outlines the controlling factors for the bearing capacity and the settlement of reinforced soil foundations. It also describes design methods which may be adopted for estimating bearing capacity of reinforced soil foundations. The vast literature available on reinforced soil foundations indicates that reinforced foundations are viable solutions for improving bearing capacity or for limiting the total and the differential settlements. Laboratory and field tests have shown that this system can be used either in lieu of, or in combination with other ground improvement techniques.

Bibliography

[1] Bindumadhava (1990) Reinforced soil technique for expansive soils, Ph.D. Thesis, Indian Institute of Science, Bangalore, India.

[2] Binquet, J. and Lee, K.L. (1975a) Bearing capacity analysis of reinforced earth mass, *Journal of Geotechnical Engg. Div.,* ASCE, 101 (12), 1241–55.

[3] Binquet, J. and Lee, K.L. (1975a) Bearing capacity analysis of reinforced earth slabs, *Journal of Geotechnical Engg. Div.,* ASCE, 101(12), 1257–76.

[4] Das, B.M. (1999) *Principles of Foundation Engineering,* Fourth Edition, PWS Publishing, Brooks/Cole Publishing Company, USA.

[5] Guido, V.A., Chang, D.K. and Sweeny, M.A. (1986) Comparison of geogrid and geotextile reinforced earth slabs, *Canadina Geotechnical Journal,* 23, 435–40

[6] Kumar, A. and Saran, S. (2001) Isolated strip fotting on reinforced şand, *Geotechnical Engineering Journal,* Southeast Asian Geotechnical Society, Vol. 32, No. 3, 177–89.

[7] Jewell, R.A. (1996) *Soil Reinforcement with Geotextiles,* Construction Industry Research and Information Association (CIRIA), Publication No.123, London.

[8] Shivashankar, R., Madhav, M.R. and Miura, N. (1993) Reinforced granular beds overlying soft clay, *Proc. 11th Southeast Asian Geotechnical Conference*, Singapore, 409–14.

[9] Singh, H.R. (1998) Bearing capacity of reinforced soil beds, Ph.D. Thesis, Indian Institute of Science, Bangalore, India.

[10] Sridharan, A. (1990) Bearing capacity improvement, *Engineering with Geosynthetics* (Eds. G.V. Rao and G.V.S.S. Raju), Tata McGraw-Hill Publishing Company Ltd, 175–96.

[11] Srinivasa Murthy, B.R. Sridharan and H.R. Singh (1993) Analysis of reinforced soil beds, *Indian Geotechnical Journal,* Vol. 23, No. 4, 447–58.

[12] Wayne, M.H., Han, J. and Akins, K. (1998) The design of geosynthetic reinforced foundations, *Proceedings of Geosynthetics in Foundation Reinforcement and Erosion Control Systems* (Eds. J. Bowders, et al.), Geotechnical Special Publication, No. 76, 1–19.

5 Design of Reinforced Soil Slopes

5.1 Introduction

An important application of soil reinforcement technique is in the improvement of stability of slopes; it is also used in the construction of steep slopes. Stability of slopes needs to be improved when the factor of safety of the existing slope is less than unity; normally codes specify that stability improvement is considered necessary when the factor of safety is less than the recommended value of 1.5. BS 8006 (1995) provides the norms for reinforcement of slopes in a number of applications (as indicated in Fig. 5.1), viz.,

(a) reinforcement of a fill in new construction,
(b) reinforcement of failed slopes,
(c) reinforcement of existing ground in cut slopes,
(d) reinforcement of an existing cut or fill slope which is marginally stable.

As indicated earlier in Chapter 2, when reinforcement is introduced, it contributes to reduction in disturbing forces and increases forces that lend stability. This consideration enables designers to improve the stability of slopes to a targeted value of factor of safety. In the literature pertaining to reinforced soils, distinction is made with regard to the inclination of the slope. For example, if the angle is close to 90°, the structure is considered a retaining wall and design methods are adopted accordingly. For a given slope, failure mechanisms such as wedge, circular/logarithmic surface, or two-part wedge mechanisms are assumed, and the magnitudes of driving forces/moments are arrived at. Resisting forces/moments due to gravity and passive resistance are evaluated and the safety factors are evaluated. In this chapter a simple illustration of the use of reinforcement to increase the stability of slope, followed by a popular method of designing reinforced soil slopes given by Jewell (1996) is presented.

5.2 General Approach

Slip failures are common in cuttings or embankments when the height of the slope reaches a critical height which depends on the soil properties such as cohesion, friction angle and bulk density. It is possible to improve the factor of safety by considering the location of critical circle (assuming a circular failure mechanism) and calculating the driving moment, restoring moments and the contribution of reinforcement in the soil to resisting moment. Figure 5.2*a* shows an embankment for which the stability needs to be improved and Fig. 5.2*b* shows the embankment with reinforcement. For equilibrium the driving moments are equated with the restoring moments, and for design the factor of safety

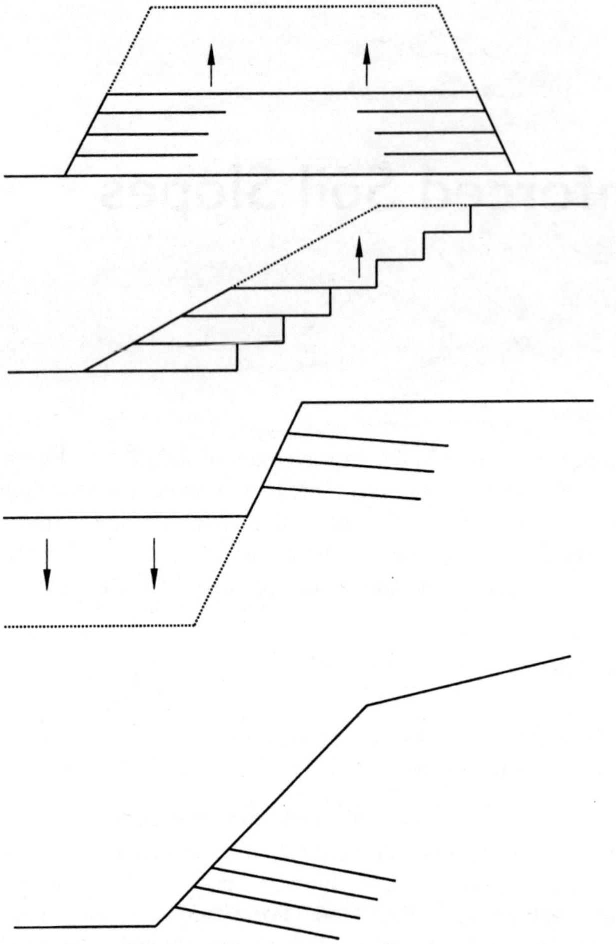

(a) Reinforcement of fill in new construction

(b) Reinforcement of failed slopes

(c) Reinforcement of existing ground in cut slopes

(d) Reinforcement to improve stability of existing structure

Fig. 5.1 *Applications of reinforcement of slopes, BS 8006 (1995)*

is obtained such that the ratio of restoring moments to the driving moments is more than the normally specified value of 1.5. If the factor of safety is less than 1.5, a reinforcement is provided taking into consideration the required additional resisting moment. This aspect is demonstrated with reference to a simple example given below. The resisting moments and the driving moments with respect to the centre of rotation.in a soil slope without reinforcement (Fig. 5.2*a*) can be calculated using the standard methods described in textbooks by Murthy (1992), Venkataramaiah (1995) and Terzaghi et al., (1996). As a typical example, assume that the driving moment (DM) and the restoring moment (RM) are given as 15000 kN m and 20000 kN m. The factor of safety (FS) is

$$\text{FS} = \frac{\text{RM}}{\text{DM}} = \frac{20000}{15000} = 1.33.$$

The resisting moment required for a factor of safety of 1.5 = 1.5× driving moment = 1.5 × 15000 = 22500 kN m.

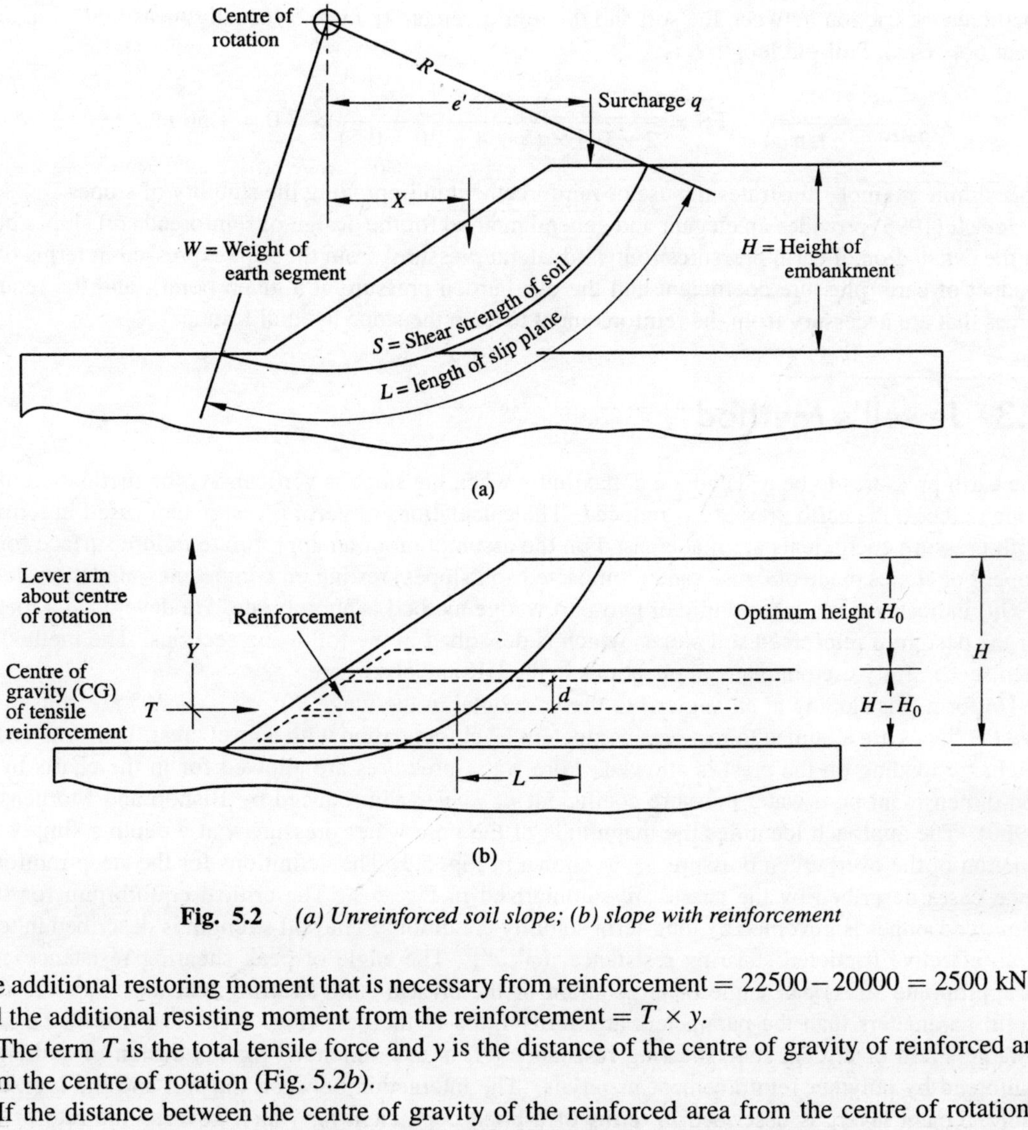

Fig. 5.2 *(a) Unreinforced soil slope; (b) slope with reinforcement*

The additional restoring moment that is necessary from reinforcement $= 22500 - 20000 = 2500$ kN m and the additional resisting moment from the reinforcement $= T \times y$.

The term T is the total tensile force and y is the distance of the centre of gravity of reinforced area from the centre of rotation (Fig. 5.2*b*).

If the distance between the centre of gravity of the reinforced area from the centre of rotation is 10 m, then, total reinforcement force $= 2500/10 = 250$ kN.

If the design strength of the reinforcement member is 35 kN/m, then the number of layers of reinforcement are $250/35 = 7.14$ or 8.

Hence the provision of 8 layers of reinforcement ensures that the factor of safety of the reinforced soil slope is 1.5. The spacing and location of the reinforcement members should be chosen in such a way that the distance between the centre of gravity of the reinforced area from the centre of rotation is maintained at 10 m as indicated in the earlier step. The length of reinforcement is governed by the overburden acting on the top of the reinforcement. Further, assume that the distance between the top level of the reinforcement and the crest of the slope is 4 m, and that the embankment soil has cohesion of 5 kN/m^2, friction angle of 26.5° and bulk unit weight of the soil of 20 kN/m^3. In addition, the coefficient of friction between the soil and the reinforcement (μ) is 0.5 (usually measured from direct shear box tests). Pull-out length L is:

$$L = \frac{Ti}{2\mu(c + \sigma \tan \phi)} \times \text{FS} = \frac{35}{2 \times 0.5 \times (5 + 4 \times 20 \times 0.5)} \times 2.0 = 1.56 \text{ m}.$$

This simple example illustrates the use of reinforcement in improving the stability of slopes.

Jewell (1996) provides an elegant and general method for the design of reinforced soil slopes based on the calculation of earth pressures (outward lateral pressures from the soil, expressed in terms of the product of earth pressure coefficient and the overburden pressure at a given point), and the required forces that are necessary from the reinforcement to keep the slope in equilibrium.

5.3 Jewell's Method

The earth pressures to be resisted are a maximum when the slope is vertical. As the inclination of the slope reduces, the earth pressure is reduced. The calculations of earth pressure expressed in terms of earth pressure coefficients are made based on the assumption of an appropriate failure surface for the slope. For slopes made of dense sand (compacted soil slopes) resting on competent foundation, Jewell (1996) indicates that a logarithmic or two-part wedge method is appropriate. He developed a method for the design of reinforced soil slopes which is described in the following sections. The method has been successfully used in many applications in the UK and elsewhere.

Uniform fills sloping at an angle β to the horizontal in the range $90° > \beta > 30°$ are considered. The fill slopes are assumed to rest on a competent level foundation with a level crest. Uniform vertical surcharge loading on the crest is allowed. Pore water pressures are allowed for in the charts by the non-dimensional pore water pressure coefficient $r_u = u/\gamma z$ introduced by Bishop and Morgenstern (1960). The approach identifies the magnitude of the pore water pressure u at a depth z simply as a function of the overburden pressure γz, as shown in Fig. 5.3. The definitions for the steep reinforced slope cases described by the charts are summarized in Fig. 5.3. The critical equilibrium for steep reinforced slopes is governed by long-term stability conditions. The soil strength is described in terms of an effective frictional shearing resistance, (ϕ', c'). The angle of peak shearing resistance ϕ'_p at an appropriate stress and angle of large strain or the critical state shearing resistance ϕ'_{cs} are more useful parameters than the parameters arbitrarily fitted to the data (Fig. 5.4). The use of the angle of large strain or critical state shearing resistance ϕ'_{cs} is recommended for the design of soil slopes reinforced by polymer reinforcement materials. The interaction between the soil and the horizontal reinforcement layers is described in terms of a bond coefficient f_b, which governs the rate of load transfer between the reinforcement and the soil (i.e., the bond length for a reinforcement layer). A separate direct sliding coefficient f_{ds} applied to the soil shearing resistance, either immediately above or below a layer of reinforcement in the reinforcement zone (Fig. 5.5), governs the resistance to shear-failure ($f_{ds} \tan \phi'$). To keep the number of charts to a minimum, Jewell (1996) prepared the charts for one value of the direct sliding coefficient, ($f_{ds} = 0.8$) which encompasses most of the practical cases safely. A simple correction can be applied where f_{ds} takes a lesser value. The calculation of these parameters is explained in subsequent sections.

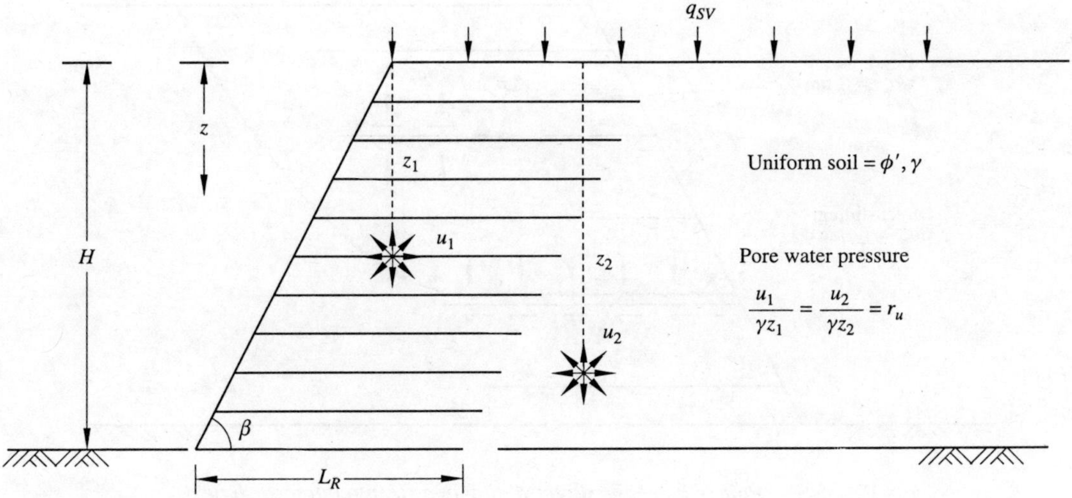

Fig. 5.3 *Definitions for steep reinforced slopes (Jewell, 1990)*

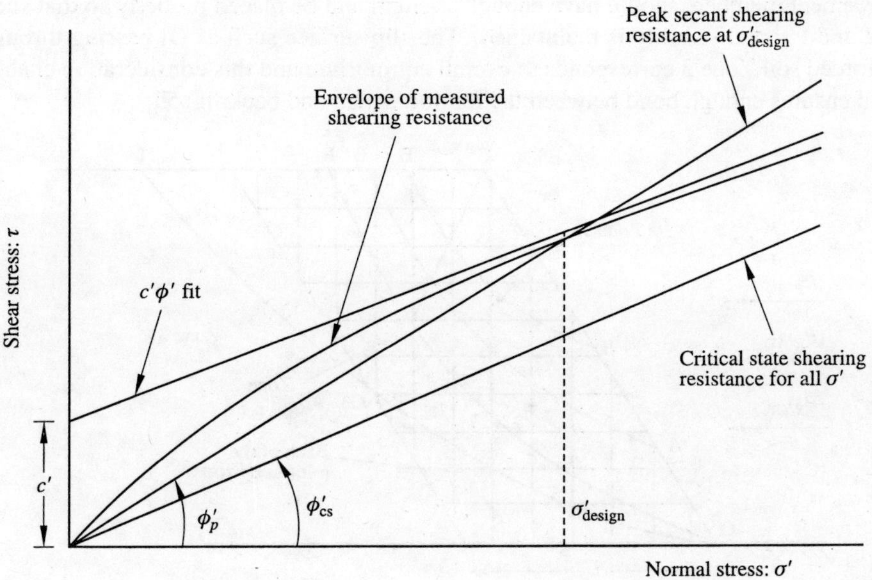

Fig. 5.4 *Envelope of shearing resistance showing a peak secant angle of friction and conventional shear strength parameters fitted to arbitrary data*

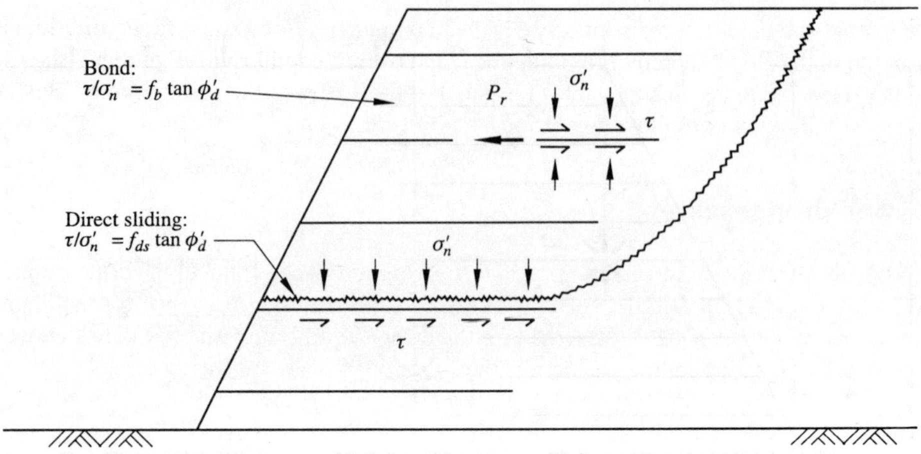

Fig. 5.5 *Role of bond and direct sliding in a steep reinforced slope*

5.4 Choice of Design Values for Parameters and Use of Charts

Three distinct zones are identified in a reinforced soil slope. In Zone 1 the reinforcement forces are required to maintain stability corresponding to the slip surfaces such as BD, AE and OF in Fig. 5.6. The reinforcement members should have enough strength and be placed properly so that such failures are avoided and internal stability is maintained. The slip surface such as OI passing through Zone 2 and unreinforced soil Zone 3 corresponds to overall equilibrium and this consideration enables overall stability and ensures enough bond between the reinforcement and backfill soil.

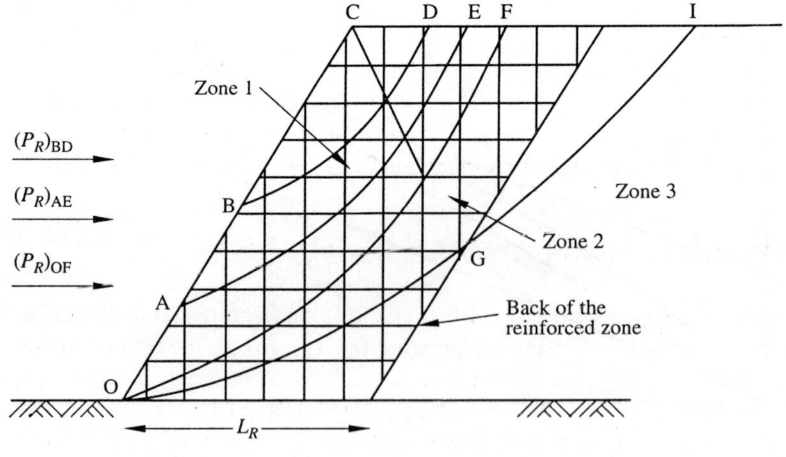

Fig. 5.6 *Zones of interest in a reinforced soil slope*

The main requirement for an economical and balanced equilibrium in a steep slope is the distribution of the maximum stress required to be supplied by the reinforcement for equilibrium in the soil. A suitable provision for reinforcement is made (with sufficient strength and spacing) so that the minimum available stress from the reinforcement exceeds the maximum required stress for equilibrium in the soil at every depth in the slope, thereby providing local and overall equilibrium. Figure 5.7 shows a typical illustration. The selection of design values for material properties of the soil, reinforcement and safety margins are explained in the following section.

5.4.1 Soil properties

It is desirable to select a design value for the soil shearing resistance equal to the critical state shearing resistance ($\phi'_d = \phi'_{cs}$). The ratio between the expected peak shearing resistance for the fill ϕ'_p and the critical state shearing resistance ϕ'_{cs} assumed for the design equilibrium, may be considered equivalent to a lumped factor of safety on the peak strength of the soil.

$$FS_s = \tan\phi'_p / \tan\phi'_{cs}.$$

Typical values of the critical state strength for granular fills are in the range $\phi'_{cs} = 30°$ to 35°, and for low plasticity clay fill $\phi'_{cs} = 20°$ to 25°. Filling and compaction should be well controlled during construction so that the maximum expected unit weight γ_d for the soil as per recommendations is achieved in the field. It may be noted that in a compacted soil, the suppression of dilatory behaviour in the soil increases the effective stresses and results in improved performance of reinforced soil structure.

Pore water pressures during the design life of the slope are more variable and less well controlled. Long term equilibrium pore water pressures as high as $r_u = 0.25$ have been recorded for grassed clay fill slopes (Vaughan et al., 1978). An appropriate design value for the pore water pressure coefficient r_u is selected by comparing the pore water pressure implied at given positions in the slope with the worst expected pore water pressure at those positions. The influence of partially wet conditions due to seepage in slopes can be examined using a value of $r_u = 0.25$.

The slope dimensions H and β may be taken equal to their expected values for design. Inclusion of a nominal vertical surcharge of about 10 kN/m^2 is often prudent in routine design to allow for some margin on over-filling and other temporary loading on the slope crest. Vertical surcharge loading is taken into account by designing the reinforcement for an equivalent slope of increased height H'. The above design values of ϕ_d, $(r_u)_d$, β and H are sufficient to determine the required earth pressure coefficient K_{req} and the required reinforcement length L_R/H from the design charts.

5.4.2 Properties of reinforcement material

The allowable force P_{all} in the reinforcement is selected to allow for: (a) conditions in the ground at the end of the design life and at the design temperature (t_d, T_d), (b) for the reinforcement material having been subject to mechanical damage during installation, and (c) subsequent action of the soil chemicals and microbiological environment. A practical procedure (Fig. 5.8) for evaluating this allowable force was described by Greenwood and Jewell (1989), and their recommended partial factors are summarised in Table 5.1. The load–extension properties supplied by manufacturers for polymer reinforcement materials are reference properties measured in the factory under standard conditions. Installation in the ground degrades these properties to lower values and this needs to be taken into consideration in the design. Besides, the loss in capacity is caused immediately by mechanical damage, and over time by environmental degradation (chemical and microbiological).

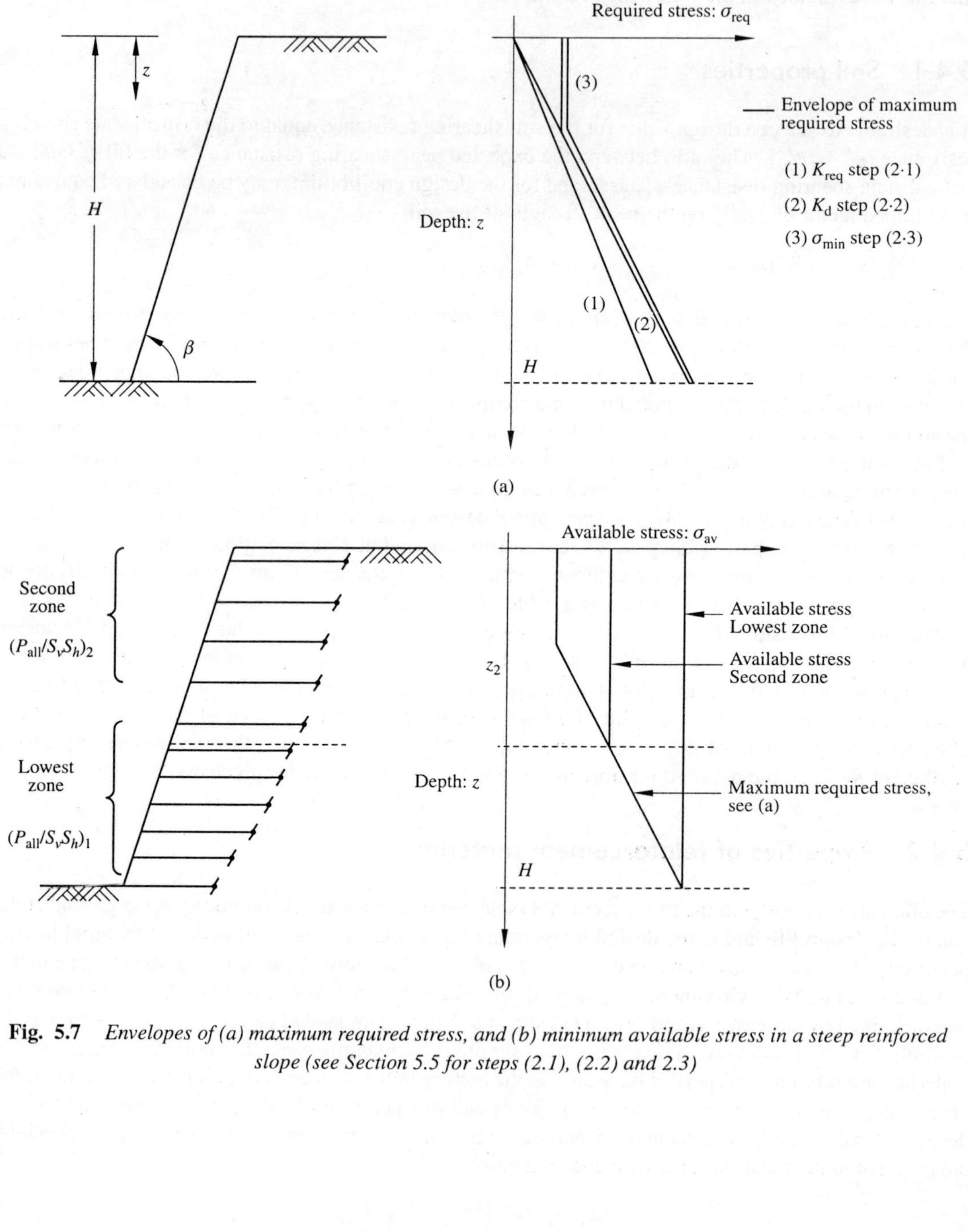

Fig. 5.7 *Envelopes of (a) maximum required stress, and (b) minimum available stress in a steep reinforced slope (see Section 5.5 for steps (2.1), (2.2) and 2.3)*

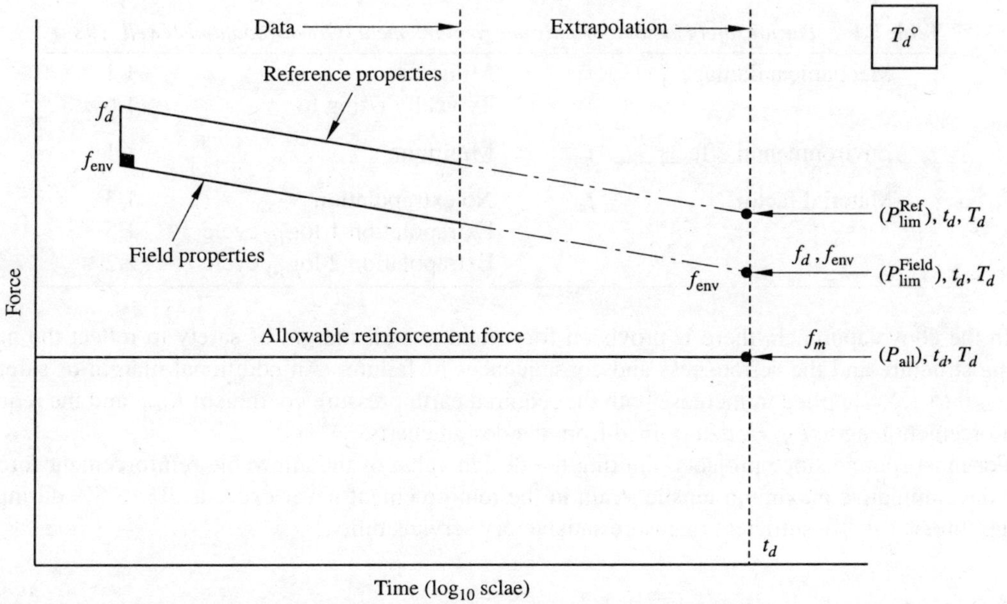

Fig. 5.8 *Reinforcement Strength characteristics at the design time and temperature (t_d and T_d) showing the degradation in the ground and the material safety factors*

The properties and degradation of polymer reinforcements are strongly governed by the time spent under load, and by the temperatures which affect the rate of reactions. The relevant reinforcement properties for the design must correspond to a load for a period of time (usually the design life of the slope), at a temperature appropriate to the field conditions (t_d, T_d). Most manufacturers use the reference rupture strength (i.e., the rupture strength indicated by the reference test data) to establish the limiting capacity of the reinforcement for extrapolation (Fig. 5.8). The corresponding field characteristic strength (the value required for design), allowing for mechanical damage and environmental degradation (f_d and f_{env} are partial factors) evaluated for the actual reinforcement product to allow for the anticipated degradation in the ground is given by

$$\left(P_{\text{lim}}^{\text{Field}}\right)_{t_d,T_d} = \frac{\left(P_{\text{lim}}^{\text{Ref}}\right)_{t_d,T_d}}{f_d f_{\text{env}}}. \tag{5.1}$$

The allowable force reduced by a material factor f_m is

$$\left(P_{\text{all}}\right)_{t_d,T_d} = \frac{\left(P_{\text{lim}}^{\text{Field}}\right)_{t_d,T_d}}{f_m}. \tag{5.2}$$

The material factor f_m allows for the uncertainty in the extrapolation of laboratory test to reach the envisaged conditions at the end of the design life t_d, at the design temperature T_d, as this is often beyond the practical range for tests (Table 5.1). Values for f_d and f_{env} recommended by manufacturers are based on tests on their reinforcement products for the entire range of construction fills and soil environments.

Table 5.1 *Partial safety factors for polymer reinforcement (Greenwood and Jewell, 1989)*

Mechanical damage	f_d	Minimum	1.1
		Typically rising to	1.6
Environmental effects	f_{env}	Minimum	1.1
Material factor	f_m	No extrapolation	1.3
		Extrapolation 1 $\log_{10}$ cycle	1.5
		Extrapolation 2 $\log_{10}$ cycle	2.2

In the above approach, there is provision for a final overall margin of safety to reflect the nature of the structure and the seriousness and consequences of failure. An additional margin of safety of $f_c = 1.0$ to 1.2 is applied to increase both the required earth pressure coefficient K_{req}, and the required reinforcement length, L_R/H, determined from the design charts.

For most routine slope projects, limiting the design value of the allowable reinforcement force so that the cumulative maximum tensile strain in the reinforcement never exceeds 3% to 5% during the design life, is usually sufficient to ensure satisfactory serviceability.

5.4.3 Interaction parameters

Direct sliding coefficient

The sliding resistance is made up of

(i) skin friction ($\tan\delta$) which acts over the proportion of the plane sliding area (a_s) that is solid reinforcement, and
(ii) The full shearing resistance of the soil ($\tan\phi'$) that acts over the remaining are of the soil $(1-a_s)$.

Thus,

$$f_{ds}\tan\phi' = a_s\tan\delta + (1-a_s)\tan\phi'$$

or

$$f_{ds} = a_s f_{sf} + (1-a_s), \tag{5.3}$$

where $f_{sf} = \tan\delta/\tan\phi'$ is the coefficient of skin friction. Figure 5.9 shows the terms of grid geometry and the mechanisms of improvement of bonding due to bearing as a result of grid apertures.

For continuous sheet reinforcement such as woven geotextiles, $a_s = 1.00$ so that the direct sliding coefficient is equivalent to the coefficient of skin friction ($f_{ds} = f_{sf}$). This can be measured in a modified direct shear test in which the soil is sheared over the reinforcement surface. For reinforcement grids the direct sliding coefficient may either be:

(a) measured in a modified shear test with soil sheared over the grid, or
(b) determined from the measured coefficient of skin friction f_{sf} for the soil on the reinforcement surface and the value of a_s from the grid geometry, substituted into the Eq. (5.3).

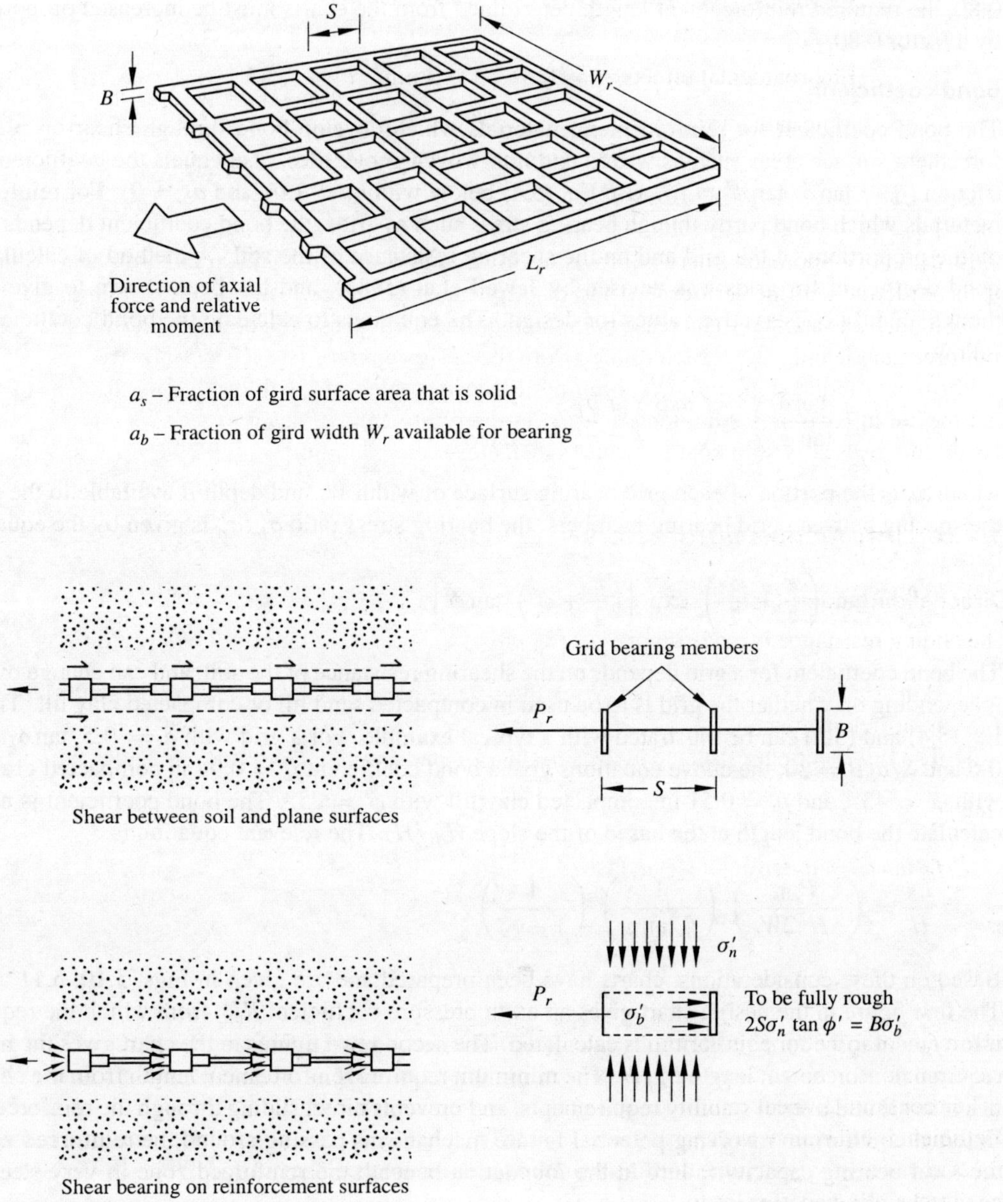

Fig. 5.9 *Terms of grid geometry and the mechanisms of improvement of bonding due to bearing as a result of grid apertures*

The latter procedure is the most practical one for slope design with reinforcement strips. If the direct sliding coefficient is less than that assumed in the derivation of the Jewell's design charts, i.e., if $f_{ds} \leq 0.80$, the required reinforcement length determined from the charts must be increased proportionately by a factor $0.80/f_{ds}$.

Bond coefficient

The bond coefficient for reinforcement materials which develop bond through shear on plane reinforcement surface areas such as woven and non-woven geotextiles, also equals the coefficient of skin friction ($f_b = \tan\delta/\tan\phi' = f_{sf}$. Use Eq. (5.4) below with $a_s = 1.00$ and $a_b = 0$). For reinforcement materials which bond partly though bearing stress such as grids, the bond coefficient depends strongly on the proportions of the grid and on the shearing resistance of the soil. A method of calculating the bond coefficient for grids was devised by Jewell et.al (1984), and has been shown to give suitable, though slightly conservative values for design. The equations to calculate the bond coefficient are as follows:

$$f_b = a_s \left(\frac{\tan\delta}{\tan\phi'}\right) + \left(\frac{a_b B}{S}\right)\left(\frac{\sigma'_b}{\sigma'_n}\right)\frac{1}{2\tan\phi'} \tag{5.4}$$

where a_b is the portion of each grid bearing surface of width W_r and depth B available to the soil. S is the spacing between grid bearing members; the bearing stress ratio σ'_b/σ'_n is given by the equation

$$\frac{\sigma'_b}{\sigma'_n} = \tan\left(\frac{\pi}{4} + \frac{\phi'}{2}\right)\exp\left[\left(\frac{\pi}{2} + \phi'\right)\tan\phi'\right]. \tag{5.5}$$

The bond coefficient for a grid depends on the shearing resistance of the soil, and can change by a factor 2 depending on whether the grid is to be used in compacted sand fill or compacted clay fill. The use of Eq. (5.4) and (5.5) can be illustrated with a typical example. For a grid with $a_s = 0.5$, $\tan\delta/\tan\phi' = 0.6$ and $S/a_b B = 20$, the above equations give a bond coefficient $f_b = 0.94$ in compacted granular fill with $\phi' = 45°$, and $f_b = 0.51$ in compacted clay fill with $\phi' = 25°$. The bond coefficient is needed to calculate the bond length at the based of the slope (L_B/H). The relevant equation is

$$\frac{L_B}{H} = \left(\frac{P_{\text{all}}}{\gamma H^2 2W_r}\right)\left(\frac{1}{f_b\tan\phi}\right)\left(\frac{1}{1-r_u}\right). \tag{5.6}$$

Based on these considerations, charts have been prepared and are given in Figs. 5.10, 5.11 and 5.12. The first figure in the design chart gives an earth pressure coefficient K_{req} from which the required reinforcement force for equilibrium is calculated. The second two figures in the chart gives the minimum required reinforcement length L_R/H. The minimum required reinforcement length from the chart satisfies internal and overall stability requirements, and prevents direct sliding through the reinforced block. External equilibrium involving potential failure mechanisms passing around the reinforced zone, and the local bearing capacity failure in the foundation beneath the reinforced zone in very steep slopes need to be checked separately.

Steep reinforced slope design charts, Jewell (1990)

Chart 1

$r_u = \frac{u}{\gamma_z} = 0.00$

Minimum required force K_{req}

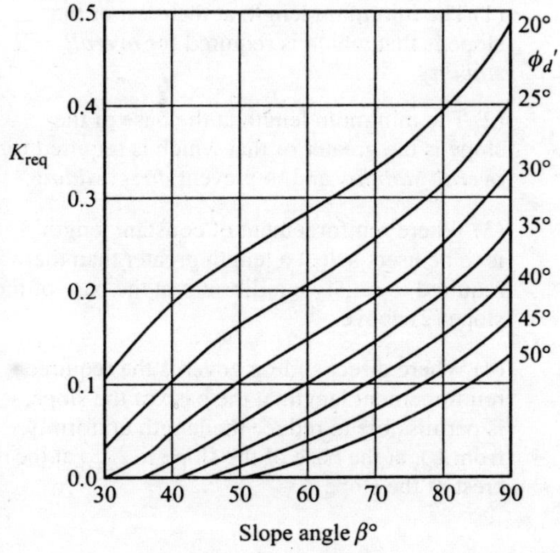

Minimum reinforcement length:

(1) The minimum length at the crest of the slope is that which is required for *overall stability*.

(2) The minimum length at the base of the slope is the greater of that which is required for *overall stability* and to prevent *direct sliding*.

(3) Where reinforcement of constant length is to be used, select a length greater than that required to satisfy equilibrium at the base of the slope (2) above.

(4) Where direct sliding governs the required reinforcement length at the base of the slope it is permissible to reduce the length uniformly from L_{ds} at the base of the slope to L_{ovrl} at the crest of the slope

Minimum required length
overall stability $(L_R/H)_{ovrl}$

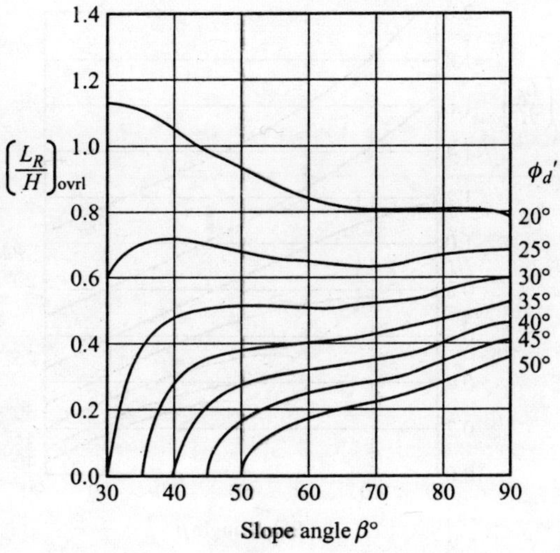

Minimum required length
direct sliding $(L_R/H)_{ds}$

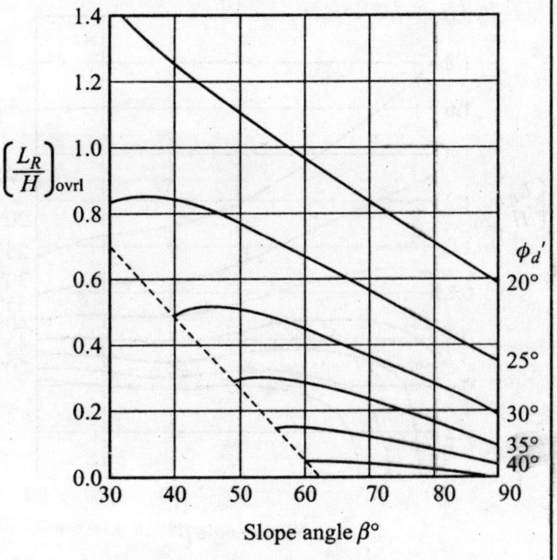

Fig. 5.10 *Design charts for reinforced soil slopes for $r_u = 0.0$ (Redrawn from Jewell, 1990)*

Steep reinforced slope design charts, Jewell (1990)	Chart 2 $r_u = \frac{u}{\gamma z} = 0.25$

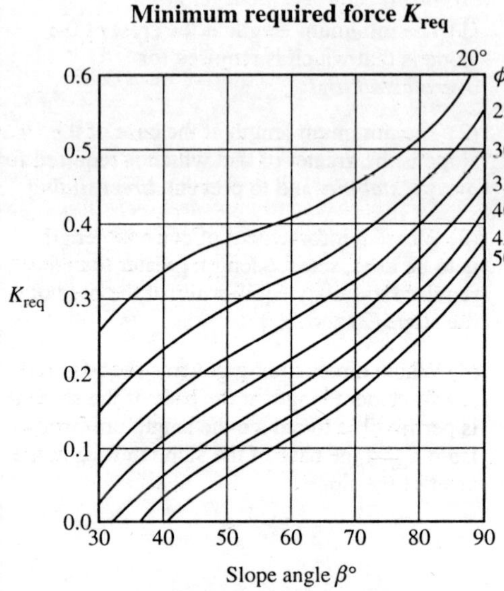

Minimum reinforcement length:

(1) The minimum length at the crest of the slope is that which is required for *overall stability*.

(2) The minimum length at the base of the slope is the greater of that which is required for *overall stability* and to prevent *direct sliding*.

(3) Where reinforcement of constant length is to be used, select a length greater than that required to satisfy equilibrium at the base of the slope (2) above.

(4) Where direct sliding governs the required reinforcement length at the base of the slope, it is permissible to reduce the length uniformly from L_{ds} at the base of the slope to L_{ovrl} at the crest of the slope

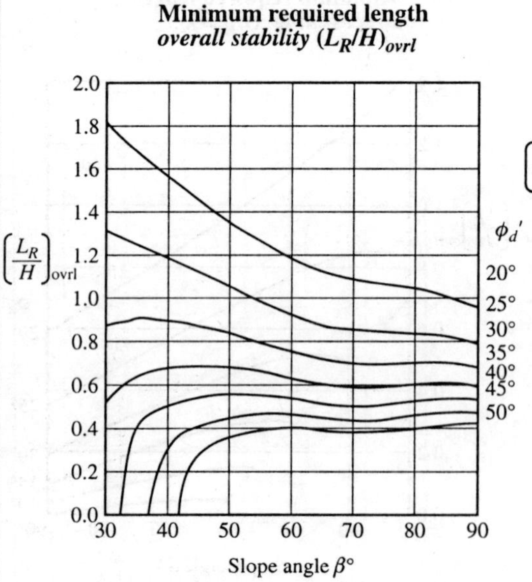

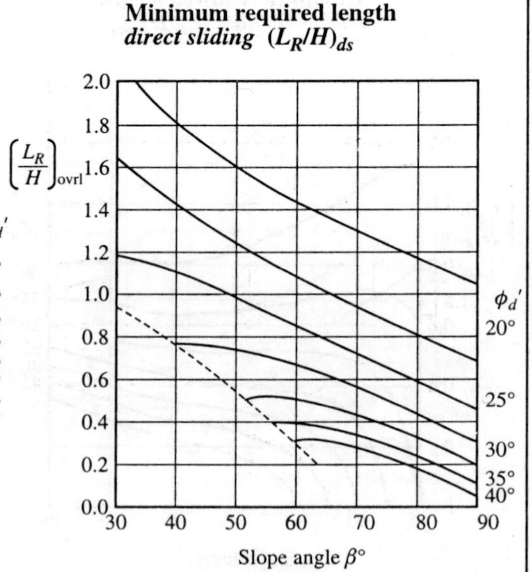

Fig. 5.11 *Design charts for reinforced soil slopes for $r_u = 0.25$ (Redrawn from Jewell, 1990)*

Steep reinforced slope design charts, (Jewell, 1990)	Chart 3 $r_u = \frac{u}{\gamma z} = 0.50$

Minimum required force K_{req}

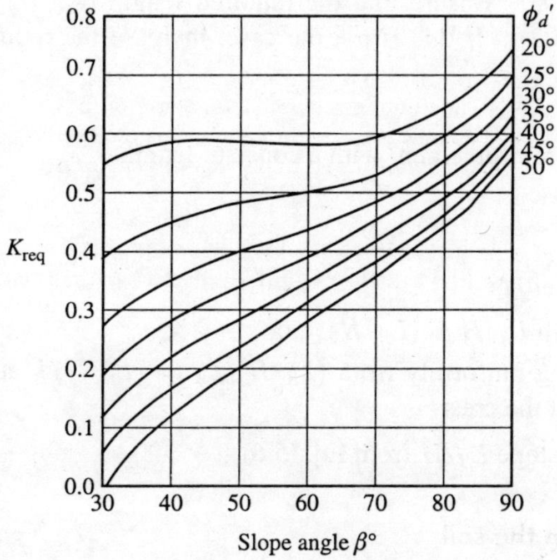

Minimum reinforcement length:

(1) The minimum length at the crest of the slope is that which is required for *overall stability*.

(2) The minimum length at the base of the slope is the greater of that which is required for *overall stability* and to prevent *direct sliding*.

(3) Where reinforcement of constant length is to be used, select a length greater than that required to satisfy equilibrium at the base of the slope (2) above.

(4) Where direct sliding governs the required reinforcement length at the base of the slope it is permissible to reduce the length uniformly from L_{ds} at the base of the slope to L_{ovrl} at the crest of the slope

Minimum required length
overall stability **$(L_R/H)_{ovrl}$**

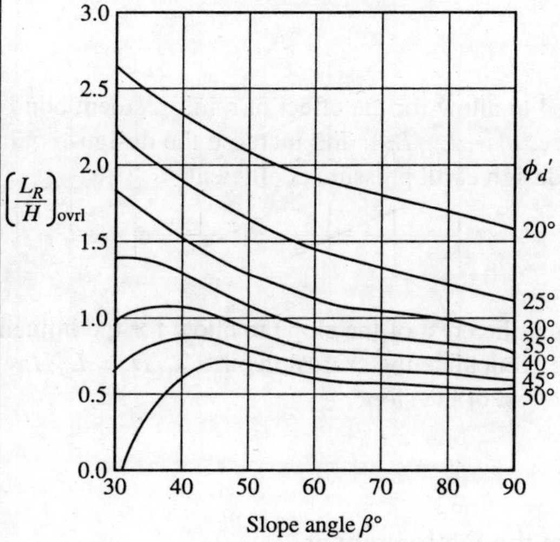

Minimum required length
direct sliding **$(L_R/H)_{ds}$**

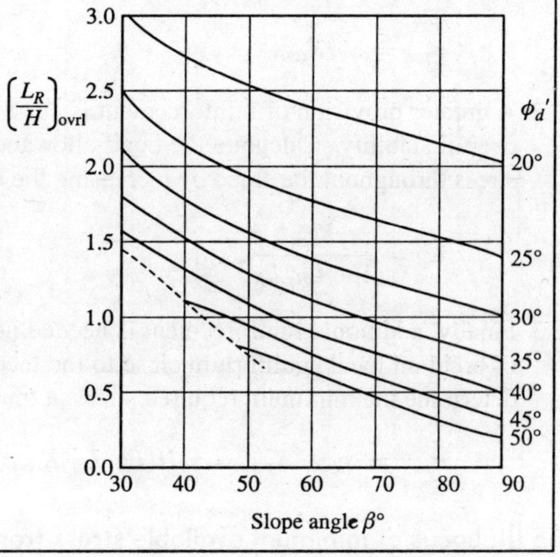

Fig. 5.12 *Design charts for reinforced soil slopes for $r_u = 0.50$ (Redrawn from Jewell, 1990)*

5.5 Steps for Simplified Design

Step I: Values from the charts

1.1 Select the design values for the parameters and determine K_{req} and L_R/H from the charts. A linear interpolation between charts is sufficient. If an overall partial factor f_c greater than unity is used, increase the values determined from the charts accordingly.

1.2 The charts have been prepared for $f_{ds} = 0.80$. Ensure that the required length $(L_R/H)_{ds}$ is valid by checking that $f_{ds} \geq 0.8$ (use Eq. (5.3)). If this is not the case, increase the required reinforcement length $(L_R/H)_{ds}$ by a factor $0.8/f_{ds}$.

1.3 Select the reinforcement length as follows.

(a) Where $(L_R/H)_{\text{ovrl}} > (L_R/H)_{ds}$, choose reinforcement with a constant length

$$L_R/H = (L_R/H)_{\text{ovrl}}.$$

(b) Where $(L_R/H)_{ds} > (L_R/H)_{\text{ovrl}}$, choose either

(i) reinforcement with constant a length $L_R/H = (L_R/H)_{ds}$, or

(ii) reinforcement with a length varying uniformly from $(L_R/H)_{\text{base}} = (L_R/H)_{ds}$ at the base to $(L_R/H)_{\text{crest}} = (L_R/H)_{\text{ovrl}}$ at the crest.

1.4 Determine the bond length at the base of the slope L_B/H from Eq. (5.6).

Step II: Envelope of maximum required stress in the soil

Construct the envelope of maximum required stress in the slope (Fig. 5.7(a)).

2.1 The basic required stress for equilibrium is determined from the depth below the slope crest (z) and the required earth pressure coefficient K_{req},

$$\sigma_{\text{req}} = \gamma_d z K_{\text{req}}.$$

2.2 A greater provision of reinforcement is needed to allow for the effect of reinforcement bond on overall stability. Calculate the bond allowance, $(1 - L_B/L_R)$, and increase the design-required stress throughout the slope by increasing the design earth pressure coefficient

$$K_d = \frac{K_{\text{req}}}{1 - L_B/L_R}.$$

2.3 Finally, additional reinforcement is needed near the crest of the slope to allow for the influence of bond on local equilibrium close to the face. Calculate the critical depth $z_{\text{crit}}/H = L_B/L_R$, to determine the minimum required stress at the crest of the slope.

$$\sigma_{\min} = \gamma_d z_{\text{crit}} K_{\text{req}} = \gamma_d H (L_B/L_R) K_{\text{req}}.$$

Step III: Locus of minimum available stress from the reinforcement

3.1 Now devise a reinforcement layout so that the minimum available stress at every depth z exceeds the envelope of maximum required stress calculated in step 2.1. The available stress, $\sigma_{\text{av}} = P_{\text{all}}/(S_v S_h)$, depends both on the reinforcement force and the spacing.

3.2 Two zones of reinforcement at constant spacing are often used. The maximum provision of reinforcement is set by the lowest layer in any zone. The lowest zone extends to the base of the slope where the provision of reinforcement must satisfy the condition:

$$P_{\text{all}}/(S_v S_h) \geq \gamma_d H K_d.$$

3.3 If the spacing is to be changed at a depth z_2 below the slope crest, the same condition must be satisfied but for the depth z_2 rather than H. This step ensures that the available forces are more than the required forces at all levels.

3.4 Determine elevations for the reinforcement layers in the slope by constructing an envelope of the available stresses, and marking the maximum depths at which the spacing may be changed. Now mark the positions of the reinforcement layers working from the base of the slope, only changing the spacing once the reinforcement layer is above the maximum depth for that zone, such as z_2 in Fig. 5.7(b).

3.5 The following practical limits to the maximum vertical spacing are suggested for design:

$$(S_v)_{\max} \leq \text{Minimum of } (H/8, 1 \text{ m})$$

Allowance for uniform vertical surcharge

A uniform vertical surcharge q_{sv} at the slope crest is allowed for by designing with an artificially greater effective height H'

$$H' = H + \frac{q_{sv}}{\gamma_d}.$$

The design proceeds exactly as before (Steps I to III above) but using H' instead of H for the slope height, and an effective depth $z' = z + q_{sv}/\gamma_d$ instead of z. The provision of reinforcement is simply terminated at the physical crest H, which occurs at an effective depth $z' = q_{sv}/\gamma_d$.

5.6 Design Example

An 8 m high slope with an angle $\beta = 70°$ and with a 15 m wide crest is needed in an urban locality. The use of polymer grid reinforcement is proposed. A granular fill is available on site. The maximum measured density of the fill is 21 kN/m^3 and the large strain or critical state angle of friction measured is in the range 22° to 28°. Hence design values $\gamma_d = 20$ kN/m^3 and $\phi'_d = 25°$ are selected from this data. The slope does not carry traffic, and hence it is considered adequate to assume zero surcharge and pore water pressure, $q_{sv} = 0$ kN/m^2 and $(r_u)_d = 0$. It may be noted that surcharge or pore pressures can be taken care of in the design. The required design life is 60 years and the temperature is the range of 25°C, giving the design conditions $t_d = 60$ years and $T_d = 20$°C.

A series of reinforcement grids with index strengths in the range 110 kN/m to 35 kN/m are available. The coefficient of skin friction for these grids is $f_{sf}(\tan\delta/\tan\phi') = 0.6$, and they all have a similar geometry with $a_s = 0.4$ and $S/a_b B = 20$. The coefficient of direct sliding can be calculated as $f_{ds} = 0.4 \times 0.6 + 0.6 = 0.24 + 0.60 = 0.84$, from Eq. (5.3). The first component comes from shear of soil on the solid grid surfaces, and the second component comes from soil-to-soil shear through the grid apertures.

The bearing stress ratio in the fill is $\sigma'_b/\sigma'_n = 4$ from Eq. (5.5), so that the bond coefficient $f_b = 0.4 \times 0.6 + 4/(20 \times 2 \tan 25) = 0.24 + 0.21 = 0.45$ from Eq. (5.4). The first component is again from surface shear, and the second component from soil bearing stress.

The reinforcement grid manufacturer recommends the following allowances for damage and environmental degradation in the reinforcement ($f_d = 1.2$ and $f_{env} = 1.1$) for the soil and environmental conditions at the project site. There are stress–rupture data for grid materials upto six years of continuous loading. Extrapolation of the material properties test data by one $\log_{10}$ cycle of time is required to reach the design conditions at the end of 60 years. Hence the appropriate material factor for design is $f_m = 1.5$ (Table 5.1). Table 5.2 shows the characteristic strength of the reinforcement grids ($t_d = 60$ years and $T_d = 20°$C) given by the manufacturer. The expected cumulative strain in the reinforcement grids for the design conditions in the slope and the allowable loads is less than 3.5% in each case. The slope is considered not to be critical and an overall partial factor $f_e = 1.0$ is selected for design.

Table 5.2 *Characteristic strength of the reinforcement grids ($t_d = 60$ years and $T_d = 20°$C)*

Index strength kN/m Eq. (5.1)	Reference strength $(P_{\text{lim}}^{\text{ref}})t_dT_d$	Field strength $(P_{\text{lim}}^{\text{Field}})t_dT_d$ Eq. (5.2)	Allowable force P_{all}
110	66	50	33
80	48	36	24
55	33	25	16.5
35	21	16	10.5

Notes:
Characteristic index and reference strength from manufacturer.
Field strength for degradation $f_d = 1.2$ and $f_{env} = 1.1$
Material factor $f_m = 1.5$ for 1 $\log_{10}$ cycle of extrapolation.

5.6.1 Design steps

Step 1.1 The basic required earth pressure coefficient $K_{req} = 0.26$ and the required reinforcement length factors $(L_R/H)_{ovrl} = 0.64$ and $(L_R/H)_{ds} = 0.57$ are determined from the charts for the design values $(r_u)_d = 0$, $\beta = 70°$ and $\phi'_d = 25°$. The above values from the charts can be used directly as the overall partial factor $f_c = 1.0$.

Step 1.2 The required length at the base of the slope is valid since $f_{ds} = 0.84$ and the charts have been prepared for a value of 0.80. The extra margin of 0.04 makes the design marginally conservative.

Step 1.3 As $(L_R/H)_{ovrl} \geq (L_R/H)_{ds}$, the minimum required reinforcement length has a constant value $L_R = 0.64 \times 8 = 5.12$ m.

Step 1.4 Design is by trial and error. Here a constant spacing $s_v = 0.5$ m is assumed throughout the slope. There are four different types of geogrids of different capacities. The maximum required stress at the base of the slope is $\sigma_{req} = K_{req}\gamma_d H = 0.26 \times 20 \times 8 = 41.6$ kN/m^2. The grid reinforcement with an index strength 80 kN/m is selected for the lowest reinforcement zone as it provides a minimum available stress $\sigma_{av} = 24/0.5 = 48$ kN/m^2. The bond length at the base of the slope for this reinforcement grid per metre width (W_r) can be calculated as follows using Eq. (5.6).

$$\frac{L_B}{H} = \left(\frac{24}{20 \times 8 \times 8 \times 2 \times 1}\right)\left(\frac{1}{0.5 \tan 25}\right)\left(\frac{1}{1-0}\right) = 0.04$$

Step 2.1 The basic required stress for equilibrium is shown in Fig. 5.13.

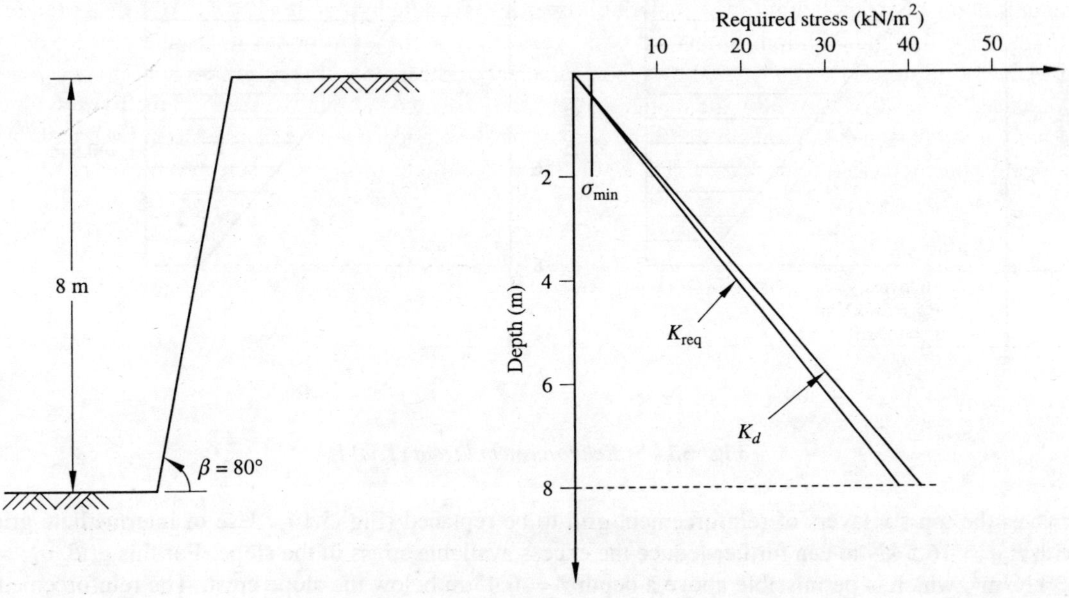

Fig. 5.13 *Basic required stress for equilibrium*

Step 2.2 For $L_R/H = 0.64$ (Step 1.3) and $L_B/H = 0.04$ (Step 1.4), $L_B/L_R = 0.063$ and the bond allowance is $(1-L_B/L_R) = 0.937$. This gives the design earth-pressure coefficient $K_d = 0.26/0.937 = 0.277$.

Step 2.3 For $L_B/L_R = 0.04$, the minimum required stress is

$$z_{\min}/H = L_B/L_R. \quad \text{Hence, } z_{\min} = 0.04H = 0.32 \text{ m}$$

$$\sigma_{\min} = K\gamma z_{\min} = 0.277 \times 20 \times 0.32 = 1.77 \text{ kN/m}^2,$$

which is very small and is indicated in Fig. 5.13.

Step 3.1 The reinforcement layout which provides a suitable locus of available stress may be devised graphically or by calculation. If a single spacing arrangement with $s_v = 0.5$ m and $P_{\text{all}} = 24$ kN m were adopted, for example, then $n = 8/0.5 = 16$ layers are required (Fig. 5.14).

Step 3.2 To investigate the use of less strong reinforcement near the crest of the slope where there is an excess available stress (as shown in Fig. 5.13), choose the weakest grid in Table 5.2 with $(P_{\text{all}})_2 = 10.5$ kN m. At the standard spacing $s_v = 0.5$ m, this grid can provide an available stress $(\sigma_{\text{av}})_2 = 21$ kN/m² which is acceptable above a depth $z_2 = 21/(0.24 \times 21) = 4.1$ m below the crest. This

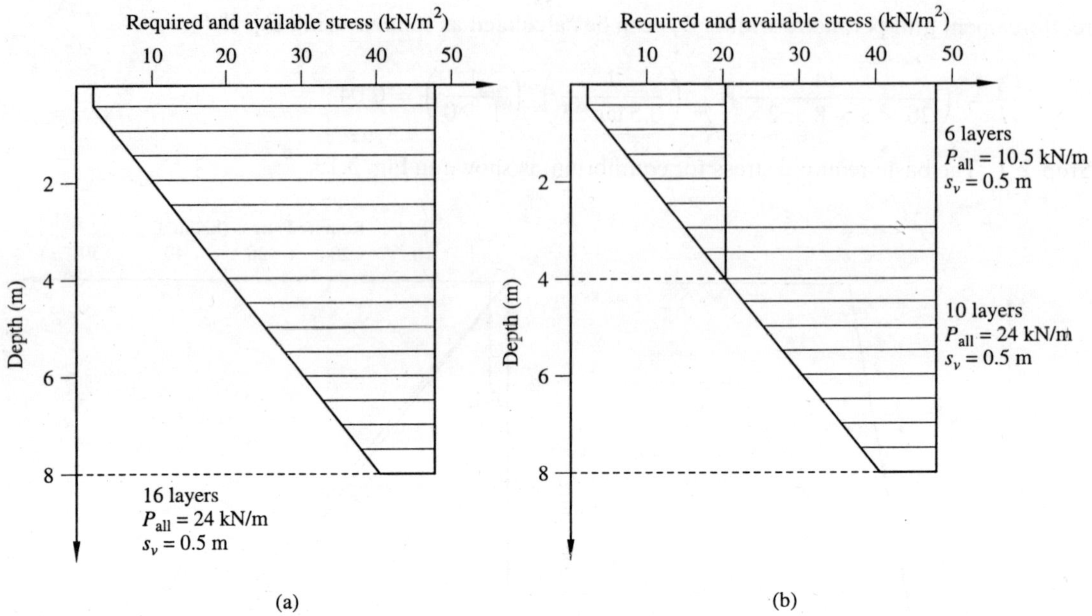

Fig. 5.14 *Reinforcement layout (Trial I)*

enables the top six layers of reinforcement grid to be replaced (Fig. 5.14). Use of intermediate grid with $P_{all} = 16.5$ kN m can further reduce the excess available stress in the slope. For this grid, $\sigma_{av} = 33$ kN/m^2, which is permissible above a depth $z = 6.45$ m below the slope crest. The reinforcement layout using all the grids has a constant spacing $s_v = 0.5$ m throughout, and five layers each of the two stronger reinforcement grids with a final six layers of the highest grid. The three alternative reinforcement designs are summarised in Table 5.3, where the relative reinforcement material costs are indicated. The optimum design will depend on individual project details. While the most sophisticated layout minimises reinforcement material costs, other cost factors, construction control and material supply arrangements can often make the simpler reinforcement layout the most attractive over all. The reinforcement layout for Design 3 is shown in Fig. 5.15.

Table 5.3 *Reinforcement design summary*

Index strength kN/m	Design 1 layers	Design 2 layers	Design 3 layers
80	16	10	5
55			5
35		6	6
Relative unit cost:	80	68	63
Saving over Design 1:	0%	15%	21%

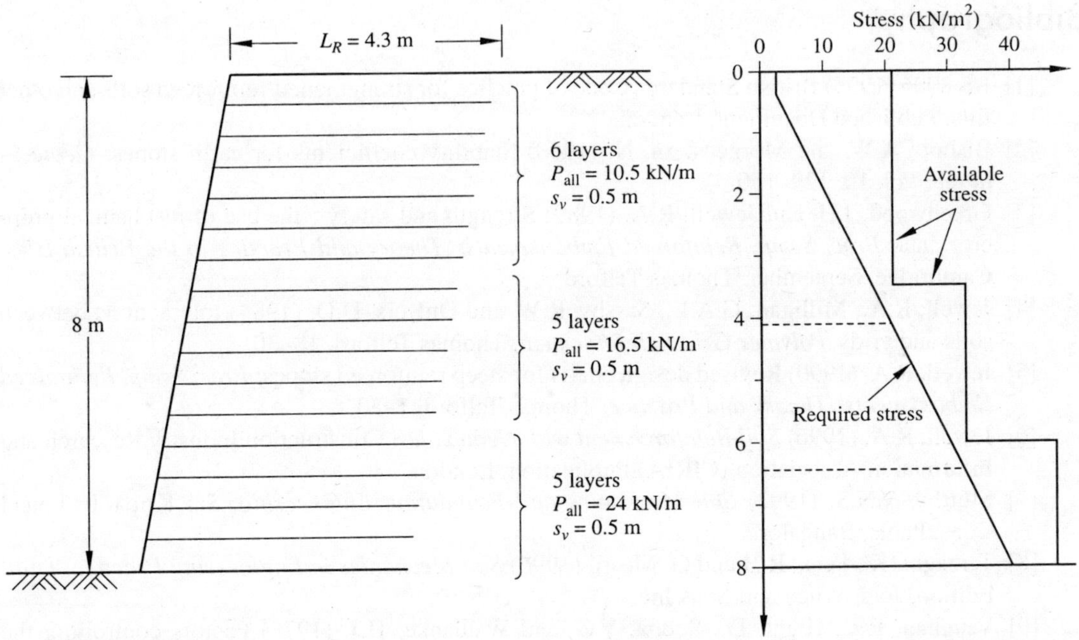

Fig. 5.15 *Reinforcement layout (Trial II)*

5.7 Role of Compaction Stresses

For very steep slopes and walls it is advisable to consider the temporary loading near the crest of the slope which may be caused by compaction. Essentially, compaction loading may require a lateral stress in the soil of the order $\sigma_{compaction} = 10\ kN/m^2$ to $20\ kN/m^2$ to maintain equilibrium. Routine design with geosynthetics often provides a sufficient quantity of reinforcement near the crest of the slope to support this magnitude of required stress. The design example described earlier is typical and a minimum available stress at the crest of $\sigma_{min} = 21\ kN/m^2$, which is more than the above value. It may also noted that compaction causes relatively short-term, transient loading compared to the value for normal design life for a slope. Polymer reinforcement materials have a higher strength capacity for short term-loading.

5.8 Concluding Remarks

This chapter presents the use of soil reinforcement in improvement of stability of slopes and the practical application of Jewell's design charts. Guidance is given on the selection of design values for soil and the reinforcement properties, and partial factors of safety. The advantages offered by the design charts and design procedure include: greater flexibility and wider range of cases, clearer and more comprehensive definition of the design allowed parameters and safety margins, and significant savings in the reinforcement material quantities.

Bibliography

[1] BS 8006 (1995) British Standard: Code of practice for strengthened/reinforced soils and other fills, Published Document, London.

[2] Bishop, A.W. and Morgenstern, N. (1960) Stability coefficients for earth slopes, *Geotechnique*, Vol. 10, 129–150.

[3] Greenwood, J.H. and Jewell, R.A. (1989) Strength and safety : the use of mechanical property data, *Proc. Symp. Reinforced Embankments: Theory and Practice in the British Isles*, Cambridge, September. Thomas Telford

[4] Jewell, R.A., Milligan, G.A.E., Sarsby, R.W. and DuBois, D.D. (1984) Interactions between soils and grids, *Polymer Grid Reinforcement*, Thomas Telford, 18–30.

[5] Jewell, R.A. (1990) Revised design charts for steep reinforced slopes, *Proc. Symp. Reinforced Embankments: Theory and Practice*, Thomas Telford, 1–30.

[6] Jewell, R.A. (1996) *Soil Reinforcement with Geotextiles*, Construction Industry Research and Information Association (CIRIA) Publication, London.

[7] Murthy, V.N.S. (1992) *Soil Mechanics and Foundation Engineering*, Sai Kripa Technical Consultants, Bangalore.

[8] Terzaghi, K., Peck, R.B and G. Mesri. (1996) *Soil Mechanics in Engineering Practice*, Third Edition, John Wiley and Sons Inc.

[9] Vaughan, P.R., Hight, D., Sodha, V.G. and Walbanke, H.J. (1978) Factors controlling the stability of clay fills in Britain, *Proc. Conf. on Clay Fills*, Institution of Civil Engineers, London.

[10] Venkatramaiah, C. (1995) *Geotechnical Engineering*, New Age International Publishers, New Delhi.

6 Design of Reinforced Soil Retaining Walls

6.1 Introduction

A number of design methods have been developed by different organizations/agencies in different geographical regions using reinforced soil techniques. These methods address issues such as availability of construction materials, particular type of reinforcing element, adaptation of the technique to local conditions, etc. As a result, a number of reinforced soil systems such as steel/geogrid reinforced retaining walls, anchored walls, soil-nailed walls, and gabion walls have evolved (Gassler and Gudehus, 1983, Jones, 1985, Brady, 1986, Leschinsky and Bodecker, 1989, Schlosser and De Buhan, 1990). Though the soil reinforcement techniques in them are quite varied, there are many common design features. In this chapter the basic mechanics of reinforced soil are reviewed and some design methods of retaining walls are explained. The calculation of required forces for equilibrium is the same for each system of reinforced soil. However, the available forces depend primarily on the reinforcement type, form, material and orientation.

6.2 Mechanics of Reinforced Soil

6.2.1 Enhancement of shear strength

Initially, reinforced soil is considered as a new composite material with the reinforcement providing an anisotropic cohesion in the direction of reinforcement. Observations on a model on full-scale reinforced-soil walls and on unit cell tests in the laboratory have shown that the reinforcement acts to alter the pattern of stresses in the soil to enable greater applied loading to be supported. Both compressive and tensile strains occur when soils are subject to shear loading, and reinforcement acts advantageously when placed in directions in which tensile strains occur. Jewell and Wroth (1987) examined the potential failure surfaces and assessed the resultant reinforcement forces acting across the failure plane using the analogy of direct shear box test. They observed the following.

- The normal component of the reinforcement force mobilizes additional frictional resistance.
- The tangential component acts to resist applied shear loading.

6.2.2 Available reinforcement force

Equilibrium is established on any potential shear surface when the combination of the mobilized shear strength and mobilized reinforcement force balance the applied shear loading. There are three limiting conditions for the magnitude of reinforcement force:

(a) the tensile strength of the reinforcement
(b) the maximum bond stresses between the reinforcement and the soil, and
(c) the compatibility of tensile strains between the reinforcement and the soil.

Towards the free end of the reinforcement, the maximum available force is likely to be governed by bond stresses, whereas towards the centre of the reinforcement, the strength or the limiting tensile strain is likely to govern. Bonds can develop between the reinforcement and the soil in both shear and bearing. Shear occurs along those surfaces parallel to the axial direction; bearing occurs on those surfaces normal to the axial direction (such as cross members in a polymer grid or ribs on a steel strip). These aspects have been dealt with in earlier chapters. The generation of bond stresses all along the reinforcement is the principal characteristic of reinforced soil which distinguishes it from traditional anchoring in which the bond is contributed by the anchor at the end of the reinforcement.

6.2.3 Stress–strain behaviour

As brought out in Chapter 2, both the soil and the reinforcement separately exhibit stress–strain response. Stress–strain models for reinforced soil are complex. This is especially true for polymeric reinforcements where the strain rate, time and temperature, all influence the performance. If the soil and reinforcement strain compatibly, then the stress–strain curves could be superimposed. Metallic reinforcements (which are relatively stiff and inextensible) can generate high forces at modest stress levels. Polymeric reinforcements (which are relatively extensible) may have to strain quite considerably in order to mobilize significant forces. This has major implications for the equilibrium of reinforced-soil walls, which derive their stability from both mobilized soil shear strength and reinforcement forces acting together. Polymer reinforced walls have to deform relatively more in order to reach equilibrium under working conditions. These aspects have been discussed in Chapter 2. It needs to be emphasized that the role of strains in both soil and reinforcement in terms of strain compatibility should be examined so that their influence on the deformation of the wall and forces mobilised in the reinforcement are understood clearly.

6.3 Components of Reinforced Soil Walls

6.3.1 Reinforcement

Many types of reinforcement are available. Generally they are made from either steel or plastic, although glass-fibre strips and other composite materials are also used. Steel and plastic reinforcements can be in the form of bars, strips, grips, meshes or (in the case of plastics) woven or non-woven fabrics. Steel reinforcement has been widely used but is susceptible to corrosion. This is an electrochemical process arising from potential differences in the soil due to the presence of metal, salts and oxygen. It is a well-understood phenomenon with considerable experience gained from steel sheet piles and buried culverts. The risk of corrosion is greater in silty or clayey soils. Protective treatments or alloying can reduce the rate of corrosion. Appropriate specifications of selection criteria for the backfill are also very useful.

In recent times, polymer reinforcements have been used in long-term retaining walls. These high-strength polymers include forms of polyethylene (grids), polypropylene (geotextiles) and polyester (geotextiles and composites). Tensile strengths of these materials vary enormously; some typical ranges are given in Table 6.1. In design, however, long-term creep strains may well prove to be more critical than tensile strength. It is therefore essential that creep data for specific products (in the form of isochronous creep curves) are made available to designers.

Table 6.1 *Typical polymer reinforcement strengths*

Polymer reinforcement type			Tensile strength (kN/m)
Geotextiles	Non-woven	Melt-bonded	3–25
		Needle-punched	7–90
		Resin-bonded	5–30
	Woven	Monofilament	20–80
		Multifilament	40–800
		Tape	8–90
	Knitted	Weft	2–5
		Warp	20–120
	Stitch-Bonded		30–1000
Geogrids	Cross-laid strips		25–200
	Punched sheets		10–110
Composites	Strips		20–150
	Bars		20–500

There is some concern over the long-term durability of polymer materials buried in soil. Vulnerability to various causes of degradation depends primarily on the polymer type, additives employed, and the manufacturing process. It seems certain that mechanical damage during construction and exposure to ultraviolet (UV) radiation prior to (or after) installation is most likely to cause a loss of tensile strength. Deterioration may be initiated by water, UV radiation, metallic ions in the soil, surface damage, and the action of solvents. Polymer behaviour may be ductile or brittle, depending on whether degradation is uniform or localized. On top of this, all degradation phenomena are accelerated by an increase in temperature.

High-density polyethylene (HDPE) frequently used in geogrids, is highly resistant to most aggressive chemicals, and can be manufactured to a high strength by mechanical stretching. The main problem with HDPE is creep (continued deformation at constant load). The major implication is in the case of reinforced soil retaining walls, where outward movement of the face may become unacceptable. The stress–strain–time relationship is complex, with high loads leading to rapid, ductile failure and low loads resulting in long-term brittle failure, occurring suddenly. Polyester fibres (used in many geotextiles and composites) are particularly susceptible to hydrolysis, in which water reacts chemically with the polymer molecules and causes loss of weight and tensile strength. The resistance of polyester to hydrolysis is compromised by exposure to UV radiation and to certain chemicals. However, these problems can be substantially overcome by encasing the polyester fibres in an inert sheath (e.g. polypropylene). Creep rarely poses a problem for polyesters at the temperatures encountered in soils.

In design, great care must be exercised if the soil is highly saline, highly organic, contains ferruginous particles, or has a high sulphate content. Strong acids or alkalis or chemical solvents (e.g. through accidental spillage on a bridge abutment) pose a serious threat to some polymers, and should be prevented from coming into contact by appropriate drainage and protection systems. UV radiation can cause very rapid degradation, and although resistance can be greatly increased by additives (e.g. carbon black) it is good practice to fully bury or cover polymer materials and keep exposure to UV during site storage to an absolute minimum. Allowance must be made for damage during installation—usually related to grading of fill and method of compaction, and expressed as a percentage of intact strength. Product-specific information should be obtained directly from manufacturers, paying careful attention to data extrapolated from short-term tests.

6.3.2 Facing

Theoretically, facing units carry little or no load and can therefore be relatively light. The primary function is to stop erosion of the fill and to provide a suitable architectural finish to the structure. In practice, the facing units will carry some lateral pressure but this will be well below the maximum pressure they are able to sustain.

Facing for reinforced soil walls can have a large variety of forms. Traditional reinforced earth makes use of cruciform concrete facing panels, although earlier walls used elliptical steel units. Grid and geotextile walls use 'wraparound' facing, porous block, brickwork, gabions, bagwork, and concrete panels (segmented and full height). Wraparound facing can be protected with shotcrete, bitumen, or vegetation. Figure 6.1 shows the common types of facings used in retaining walls.

One crucial point of detail concerns the nature of connection between the reinforcement and the facing. After construction of a reinforced soil wall, the fill will experience self-weight compression and settlement, taking the reinforcement with it. If a rigid facing is employed, differential movement will occur at the connections and may cause shear failure of the reinforcement. This needs to be considered in the design of connections.

6.3.3 Backfill

For reinforced walls, selected granular fill is preferable. Typical specifications for backfill soils were already discussed in Chapter 2 (See Tables 2.2, 2.3 and 2.4), wherein the factors influencing the performance and behaviour of reinforced soils were presented. Waste materials such as pulverized fuel ash (PFA) in combination with non-metallic reinforcement may be used. The aggressiveness of a backfill is defined with reference to its resistivity, pH, chloride content, sulphate, redox potential, etc. In practice, only a limited range of soils is likely to be used in reinforced soil walls. The trade-off between the economics of using local material and the technical superiority of imported granular fill needs to be examined.

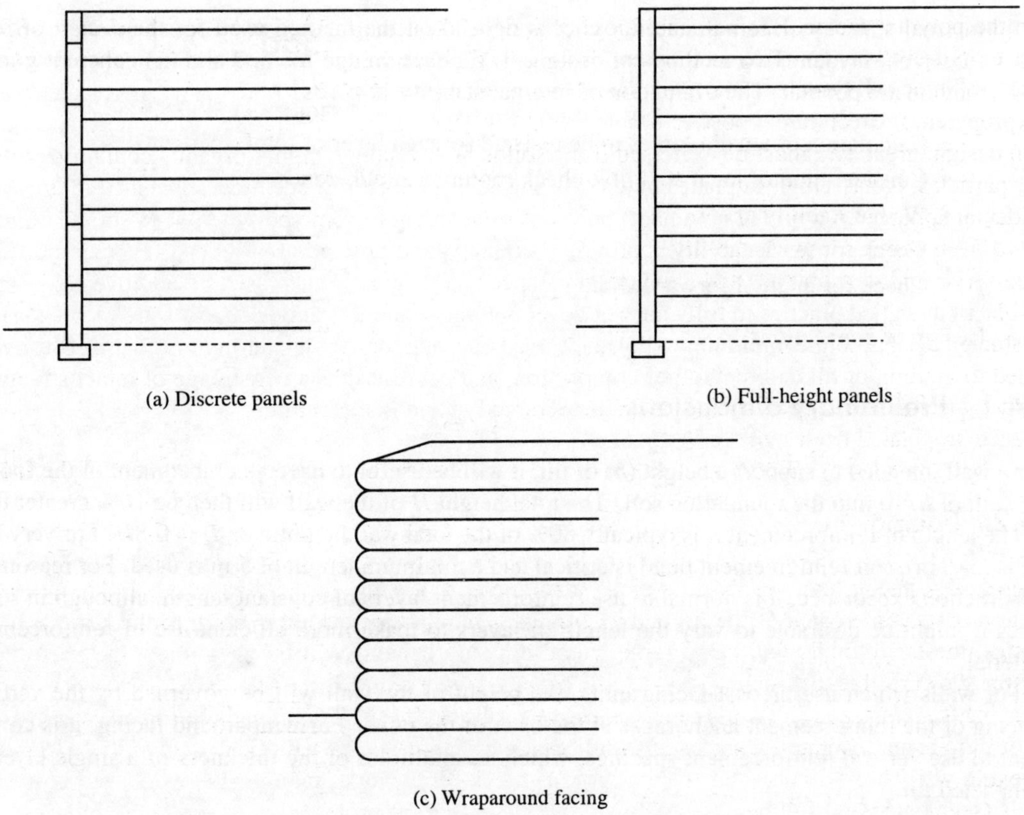

Fig. 6.1 *Different types of facings*

6.4 Principles of Design

The reinforced soil structure must have an adequate margin of safety against outward sliding, overturning, bearing failure, and deep-seated failure, without any one layer in the reinforced zone becoming overstressed. Reinforced soil walls must be checked with respect to two forms of stability:

(a) External stability

(b) Internal stability

External stability governs reinforcement length and internal stability governs reinforcement spacing. General design procedure laid down in BS 8006 (1995) consists of the following steps/checks.

1. Initial size of the structure
2. External stability check
3. Selection of the type of reinforcement

4. Internal stability: Internal stability checks depend on the method used for the design of reinforcement layout. Two methods of design, (i) tie-back wedge method and (ii) coherent gravity method are popular. The evaluation of internal stability involves:
   - Calculation of tensile forces to be resisted by each layer of reinforcement
   - Consideration of local stability: check rupture and adherence
   - Wedge stability check
   - Check for serviceability
   - Check for internal forward sliding
5. Design of connections

6.4.1 Preliminary dimensions

For a wall intended to support a height (h) of fill, it will be useful to have an embedment of the foot of the wall of $h/10$ into the foundation soil. The total height H of the wall will then be 10% greater than h. The length of reinforcement L is typically 80% of the total wall height, i.e., $L = 0.8H$. For very low walls (3–6 m), soil/reinforcement bond is critical and a minimum length of 5 m is used. For reasons of construction expedience, it is normal to use reinforcement layers of constant length, although in some cases it might be desirable to vary the length of layers to make more efficient use of reinforcement material.

For walls which use precast facing units, the height of the wall will be governed by the vertical spacing of the reinforcement anchorages at the back of the units. For wraparound facing, it is convenient to use vertical reinforcement spacings, which are multiples of the thickness of a single layer of compacted fill.

6.4.2 Factors of safety

Similar to conventional retaining wall design, global factors of safety have generally been employed for reinforced soil wall designs which normally proceed on the basis of expected values for loading (soil weight, surcharge, line loads, etc.) and resistances (soil shear strength, permissible reinforcement stress, etc.) For each possible failure mode, both external and internal, a specified factor of safety must be achieved.

This is essentially a working stress method of design. In recent times, design methods based on limit states have been formulated. It is realized that a partial factor approach to designing reinforced soil walls/reinforced soil structures should be adopted. This is particularly true of the analysis of internal stability due to the role of material properties of the reinforcement. Where polymer materials are concerned (in all their different types and geometrical forms), the picture is complicated and a partial factor approach seems to offer the only rational basis. The British Standard for reinforced soil construction, BS 8006 (1995), advocates the use of partial factors in design. There are a number of different component factors comprising the partial material factor, viz.

- Manufacture: covering deviations from the characteristic value of tensile strength (fm_{11})
- Extrapolation: taking account of the uncertainty in assessing long-term performance from short-term data (fm_{12})

- Susceptibility to damage: covering strength reductions arising from construction damage (fm_{22})
- Environment: to take account of the degradation in the ground over the design life (fm_{21}).

The following minimum values for the component factors can be used in the absence of any guidelines:

Partial factor	fm_{11}	fm_{12}	fm_{22}	fm_{21}
Minimum value	1.05	$\log(t_d/t_t)$	1.08	1.08

t_d is the design life, and t_t is the time over which tests have been conducted. Some manufacturers of proprietary polymer reinforcements have published partial material factors which they recommended for use with their range of products.

6.4.3 External stability

The external stability of a reinforced soil is assessed in the same way as that of a conventional gravity retaining wall. Considering the reinforced zone as a monolith, the mechanisms of overall failure which must be checked are shown in Fig. 6.2. It is assumed that the wall fill and the underlying foundation soil have different properties. When this is the case, values of coefficients of earth pressure should relate to the appropriate soil zones.

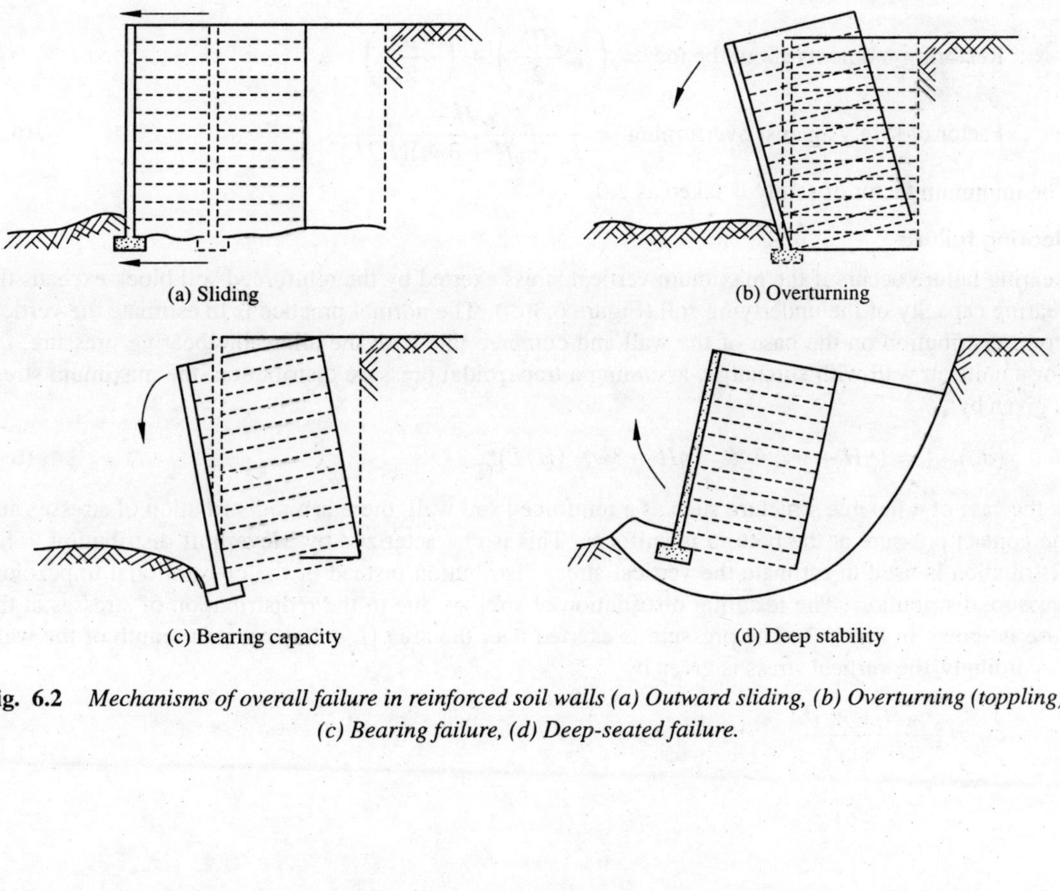

Fig. 6.2 *Mechanisms of overall failure in reinforced soil walls (a) Outward sliding, (b) Overturning (toppling), (c) Bearing failure, (d) Deep-seated failure.*

Outward sliding

Outward sliding is initiated by the thrust of the unreinforced backfill, and is most likely to occur on a plane just above or below the lowest level of reinforcement (Fig. 6.2(a)). This is because the coefficient of friction will generally be lower between soil and reinforcement than soil and soil. The factor of safety against outward sliding is based on simple of forces equilibrium, as a ratio of sliding forces to resisting forces given by

$$FS = \frac{\mu\left(\gamma_w HL + w_s L\right)}{\left(K_{ab}\gamma_b H^2/2 + K_{ab} w_s H\right)}. \tag{6.1}$$

A minimum value of 2.0 is specified. Alternatively, the minimum reinforcement length can be expressed as:

$$L_{\min} \geq \frac{FS K_{ab}\left(\gamma_w H + 2w_s\right) H}{2\mu\left(\gamma_w H + w_s\right)} \tag{6.2}$$

Overturning

Overturning is initiated by the thrust of the unreinforced backfill, causing the reinforced block to topple forward (Fig. 6.2(b)). The factor of safety is calculated from the overturning and restoring moments above the toe of the wall.

$$\text{Overturning moments about the toe} = \left(k_{ab}\gamma_b\frac{H^3}{6} + k_{ab}\frac{w_s H^2}{2}\right).$$

$$\text{Restoring moments about the toe} = \left(\gamma_w\frac{HL^2}{2}\right) + \left(\frac{w_s L^2}{2}\right).$$

$$\text{Factor of safety against overturning} = \frac{3(\gamma_w H + w_s)}{k_{ab}(\gamma_b H + 3w_s)(H/L)^2}. \tag{6.3}$$

The minimum factor of safety is taken as 2.0.

Bearing failure

Bearing failure occurs if the maximum vertical stress exerted by the reinforced soil block exceeds the bearing capacity of the underlying soil (Figure 6.3(c)). The normal practice is to estimate the vertical stress distribution on the base of the wall and compare this with the allowable bearing pressure, q_a. For a uniform wall with surcharge, assuming a trapezoidal pressure distribution, the maximum stress is given by

$$(\sigma_v)_{\max} = (\gamma H + w_s) + K_a\,(\gamma H + 3w_s)\,(H/L)^2. \tag{6.4}$$

In the case of a flexible structure such as a reinforced soil wall, there is a redistribution of stresses and the contact pressure at the bottom is uniform. This is characterized by Meyerhoff distribution. This distribution is used to estimate the vertical stress distribution instead of the conventional trapezoidal pressure distribution. The resulting distribution of stresses due to the redistribution of stresses at the base is shown in Fig. 6.3. The pressure is exerted over the area $(L - 2e)$ per unit length of the wall. Accordingly, the vertical stress is given by

$$\sigma_v = \frac{(\gamma_w H + w_s)L}{(L - 2e)},$$

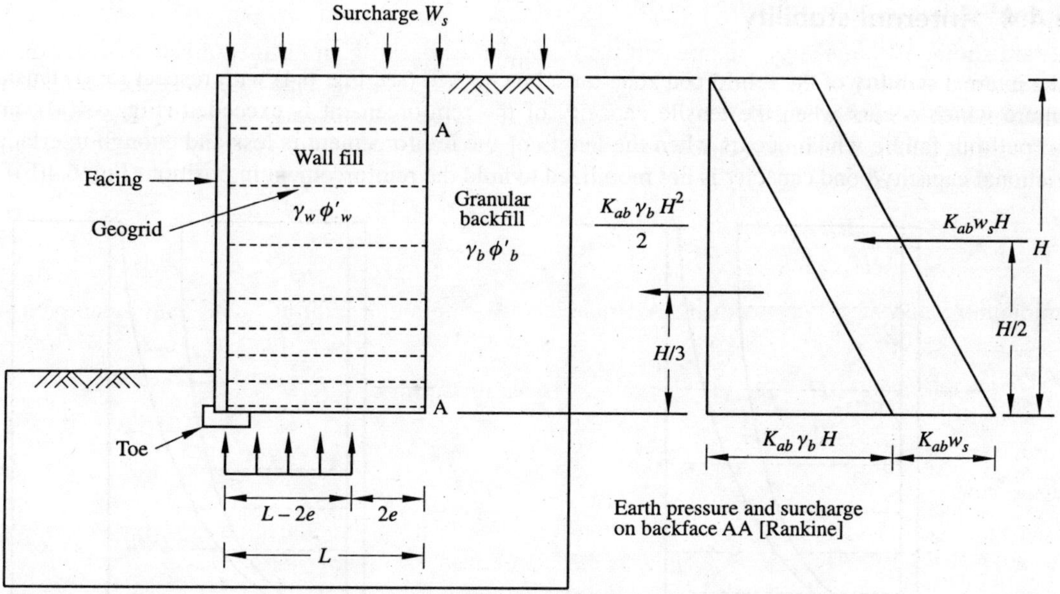

Fig. 6.3 *Forces acting on the retaining wall for consideration of external stability*

where e is the eccentricity with respect to the centre-line and the line of action of the resultant and is calculated as ratio of overturning moment to the vertical load.

$$e = \frac{(k_{ab}\gamma_b \frac{H^3}{6} + k_{ab}w_s\frac{H^2}{2})}{(\gamma_w HL + w_s L)} = \frac{k_{ab}H^2(\gamma_b H + 3w_s)}{6L(\gamma_w H + w_s)}.$$

Hence the vertical stress σ_v is

$$(\sigma_v)_{\max} = \frac{(\gamma_w H + w_s)}{1 - \dfrac{k_{ab}(\gamma_b H + 3w_s)}{3(\gamma_w H + w_s)}}. \tag{6.5}$$

The allowable bearing pressure may be estimated using bearing capacity theories. If bearing capacity is inadequate, the designer may consider using a reinforced soil foundation under the wall. This involves excavation of the subsoil to a particular depth, and replacement by layers of granular fill and geotextile/geogrid. Alternatively, a cellular mattress can be constructed with interlocking vertical sheets of geogrid, backfilled with sand and gravel.

Deep-seated failure

The analysis of deep-seated failure proceeds along lines identical to those of conventional gravity structures (Fig. 6.2(d)). An appropriate stability calculation (using circular or non-circular mechanisms) should be carried out to ensure an overall factor of safety of more than 1.5.

6.4.4 Internal stability

The internal stability of the reinforced zone must be checked (see Fig. 6.4) with respect to: (i) tension failure which occurs when the tensile capacity of the reinforcement is exceeded (Fig. 6.4(a)), and (ii) pull-out failure which occurs when the length of the reinforcement is less and enough interfacial frictional capacity/bond capacity is not mobilized to hold the reinforcement in position (Fig. 6.4(b)).

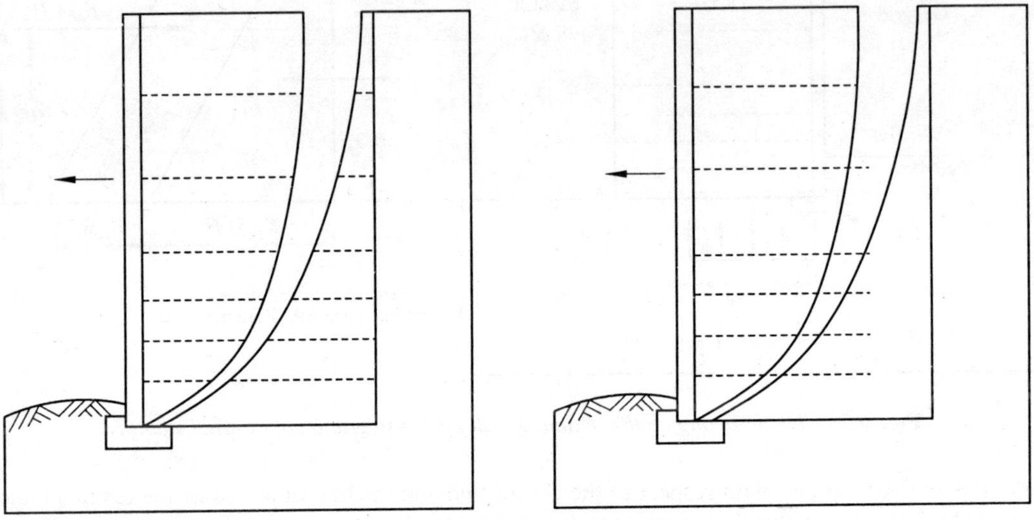

Fig. 6.4 *Mechanisms of internal failure in reinforced soil walls*

Tension failure is checked for each layer taking into account the self-weight of the fill, uniform surcharge, vertical and horizontal line loads on the crest, and bending moments caused by external loading.

Pull-out failure is checked by considering both the pull-out capacity of individual layers and the equilibrium of planar wedge mechanisms through the reinforced zone. For each layer, bond lengths beyond the point of maximum tension are assessed. Bond or anchorage lengths beyond critical wedges must be sufficient to prevent pull-out.

Analytical methods

To check the internal stability, analytical methods mostly based on Coulomb and Rankine theories are used. The two most often used in design are (i) tie-back wedge analysis, and (ii) coherent gravity analysis.

Tie-back wedge analysis assumes that the wall will deform in service in such a way as to cause the face to rotate outwards about the toe (Fig. 6.5(a)). Active conditions are assumed throughout the reinforced zone, with a constant value of earth-pressure coefficient K_a. A Coulomb-type wedge inclined at $(45 + \phi'/2)$ with respect to the horizontal is adopted as the critical failure mode, and is assumed to coincide with the locus of maximum reinforcement tension.

The coherent gravity analysis (Fig. 6.5(b)) is a semi-empirical method based on earth-pressure theory and measurements taken on many models and full-scale walls. Wall deformation is assumed to be

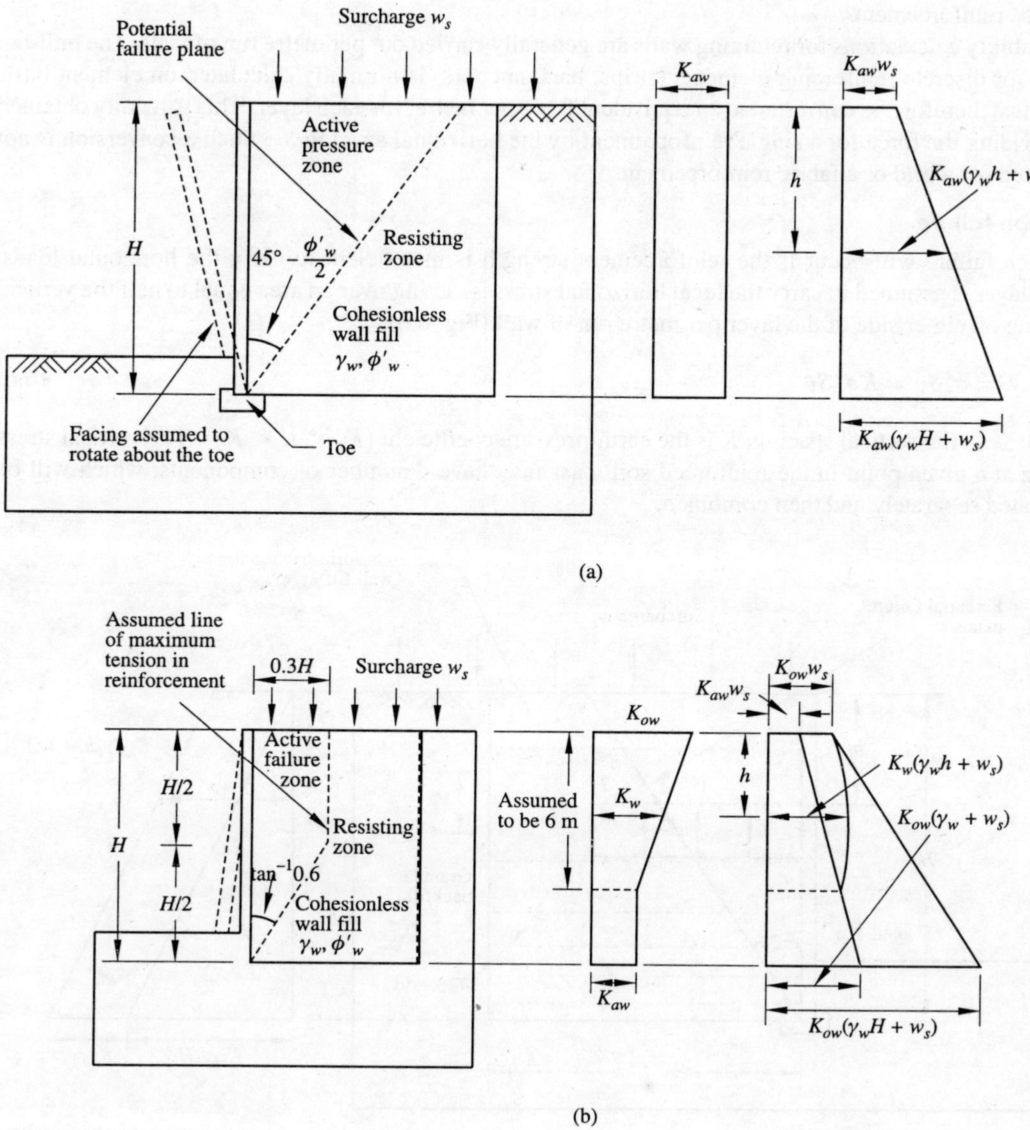

Fig. 6.5 *(a) Assumptions in tie-back wedge analysis (b) Assumptions in the coherent gravity analysis*

an outward rotation of the face about the top. A line of maximum reinforcement tension is assumed to divide the reinforced mass into an active zone and a resistant zone. A variable earth-pressure coefficient is assumed, having a value K_0 at the top, which reduces to K_a at 6 m depth and below. In addition, empirical rules have been given for the variation of coefficient of friction μ between the soil and the reinforcement.

Stability calculations for retaining walls are generally carried out per metre run of wall. The pull-out force for discrete reinforcing elements (strips, bars, anchors) is normally calculated on element basis and must therefore be converted to an equivalent force per metre, for each layer. This is readily obtained by dividing the force for a single reinforcement by the horizontal spacing S_h. Such a conversion is not required for a grid or a fabric reinforcement.

Tension failure

Tension failure will occur if the reinforcement strength is insufficient to carry the horizontal loads. Each layer is assumed to carry the local horizontal stresses, acting over an area equal to half the vertical spacing on either side of the layer per metre run of wall (Fig. 6.6).

$$T = \sigma'_h S_V = K\sigma'_v S_V, \tag{6.6}$$

where S_V is the vertical spacing, K is the earth-pressure coefficient ($K_a < K < K_0$). The vertical stress acting at a given point in the reinforced soil mass may have a number of components, which will be evaluated separately and then combined.

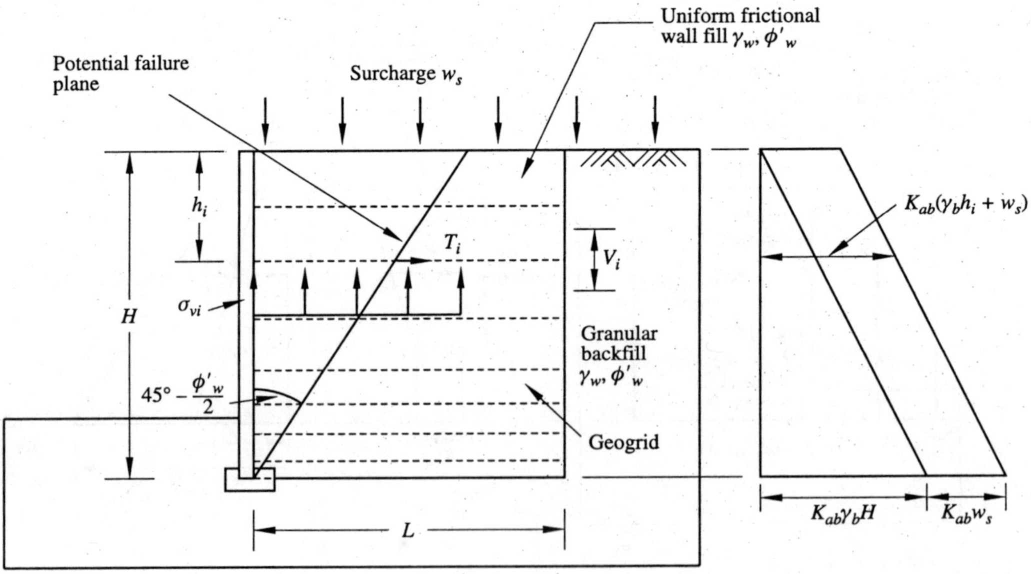

Fig. 6.6 *Calculation of tensile forces considering spacing as well as stress distridution*

Self-weight and surcharge: Equations (6.4) and (6.5), presented earlier for maximum vertical stress at the base of the reinforced zone, can be used to determine σ_{vi} at any level in the wall. The depth z

below the top of the wall is substituted for the height H. If trapezoidal distribution is used (Fig. 6.6), the vertical stress is

$$\sigma_{vi} = (\gamma z + w_S) + K_a(\gamma z + 3w_S)(z/L)^2. \tag{6.7}$$

Vertical line loads: These may arise if the wall is being used as a bridge abutment to support a bank seat. Loads are assumed to disperse with depth at a conventional spread of 1:2 (horizontal: vertical) as shown in Fig. 6.7. The vertical stress at some depth is given by

$$\sigma_v = F_v/B. \tag{6.8}$$

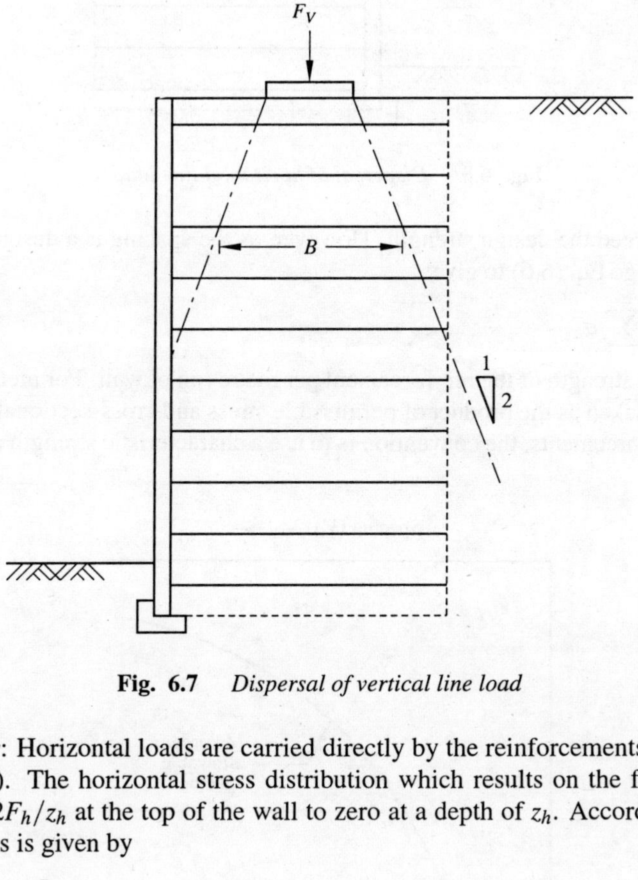

Fig. 6.7 *Dispersal of vertical line load*

Horizontal line loads: Horizontal loads are carried directly by the reinforcements in the upper portion of the wall (Fig. 6.8). The horizontal stress distribution which results on the front face of the wall varies linearly from $2F_h/z_h$ at the top of the wall to zero at a depth of z_h. Accordingly, at some depth z, the horizontal stress is given by

$$\sigma_h = \frac{2F_h}{z_h}\left[1 - \frac{z}{z_h}\right]. \tag{6.9}$$

Total tension: The sum of vertical stresses from Eqns. (6.5) to (6.8) multiplied by the earth pressure coefficient and the horizontal stress from Eq. (6.9) gives the total horizontal stresses ($\sum \sigma_h$) to be resisted by the reinforcement.

The vertical spacing S_V in the Eq (6.6) can be calculated for different depths upto the full height of the wall as shown in Fig. 6.9. This is used to determine the layout of the grids so that tension within

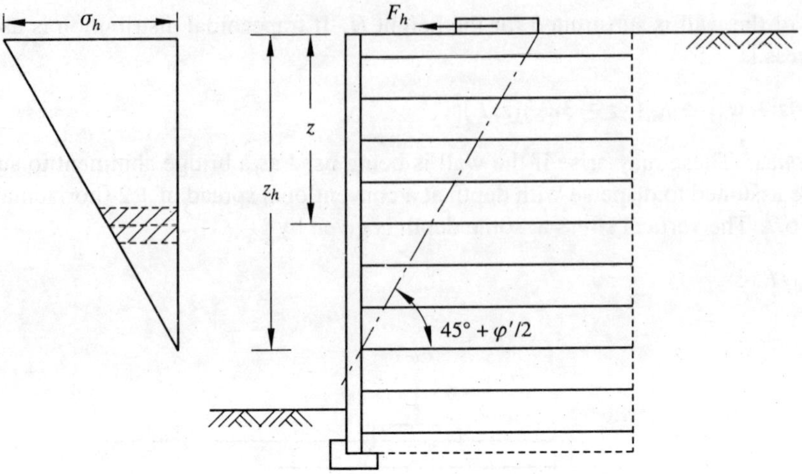

Fig. 6.8 *Dispersal of horizontal line load*

each grid does not exceed the design strength. However, as the spacing is a design variable, it may be more useful to rearrange Eq. (6.6) to give:

$$(S_V)_{\max} = P_d / \sum \sigma_h, \tag{6.10}$$

where P_d is the design strength of the reinforcement per metre run of wall. For metallic reinforcements, the design strength is taken as the product of permissible stress and cross-sectional area of the element. For non-metallic reinforcements, the convention is to use a characteristic strength reduced by a number of partial factors.

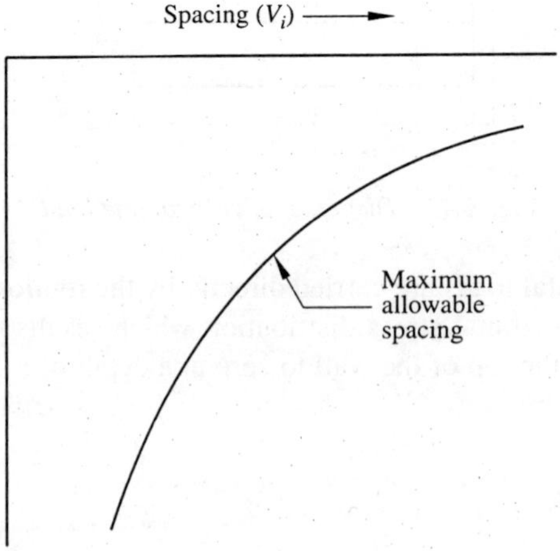

Fig. 6.9 *Variation of tensile force versus depth*

Equation (6.10) is plotted against depth below the surface to give a useful indication of the required spacing. Practical reinforcement spacing may be governed by facing connections or by compaction layer thickness. A constant vertical spacing is the most convenient one from a construction viewpoint, but in larger walls it may be desirable (and more efficient) to vary the spacing in two or more zones.

Wedge/Pull-out failure

Apart from checking for tension, it is necessary to consider the possibilities of inclined planes passing through the wall, forming unstable wedges of soil bounded by the front face of the wall, the top ground surface and the potential failure plane. The following assumptions are made to examine the wedge stability.

(a) Each wedge behaves as a rigid body.
(b) Friction between the facing and the fill is ignored.
(c) No potential failure plane passes through the contact area under a bridge abutment bank seat or similar support.
(d) The forces such as
- self-weight of the fill in the wedge
- uniformly distributed surcharge (w_s)
- vertical loading (S_1)
- horizontal loading (F_1)
- resultant force on the potential failure plane (R), and
- total tension (T) in the grids intercepted by the potential failure plane

need be considered. These forces are indicated in Fig. 6.10.

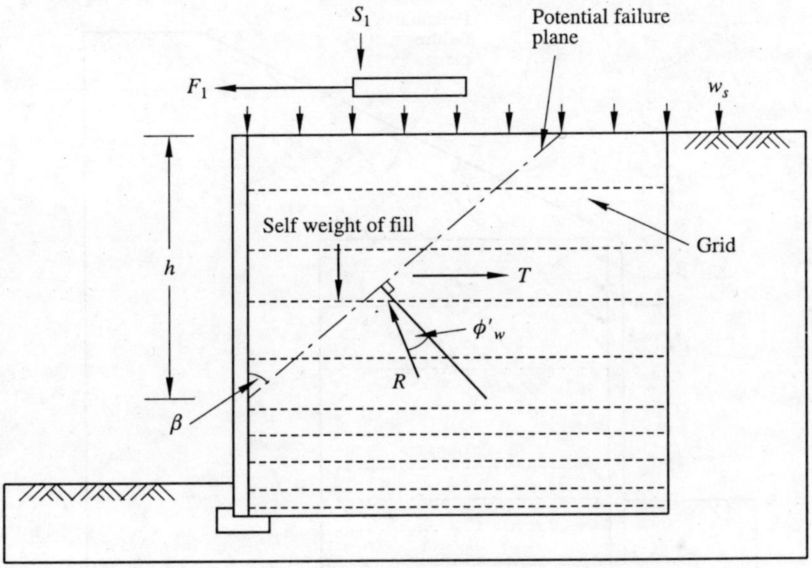

Fig. 6.10 *Forces required for wedge stability assessment*

A selection of potential failure planes should be investigated for each of the typical points indicated in Fig. 6.11. The forces acting on each wedge should be resolved horizontally and vertically (mutually

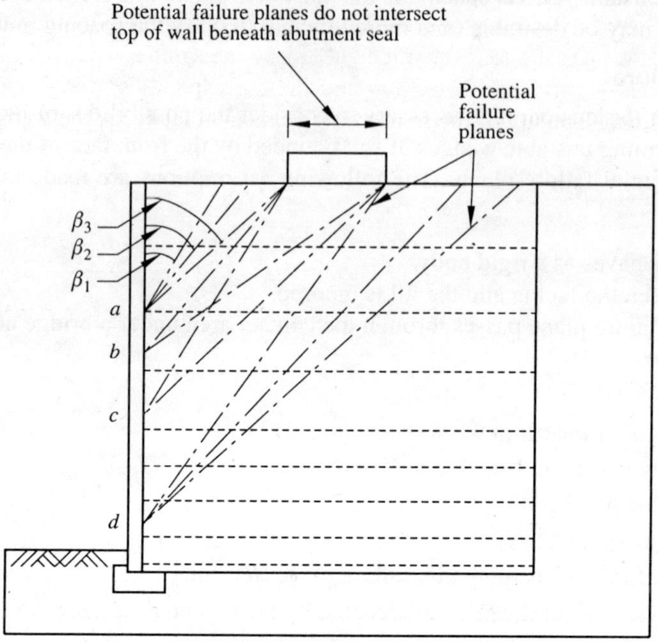

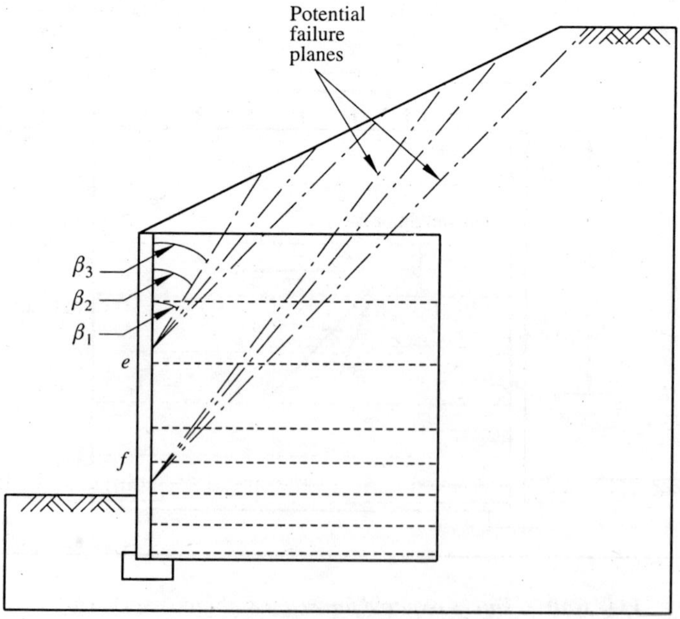

Fig. 6.11 *Potential failure surfaces to be investigated for wedge failure*

perpendicular directions). Since the forces are assumed to be in equilibrium, the two equations need to be solved simultaneously to give the tensile force (T) to be supplied by the reinforcing element. For each of the typical wedges the tension value should be determined and maximum value established (Fig. 6.12). The maximum value of tension and the corresponding wedge angle should be used to determine the frictional capacity of the reinforcing elements that keeps the wedge in equilibrium.

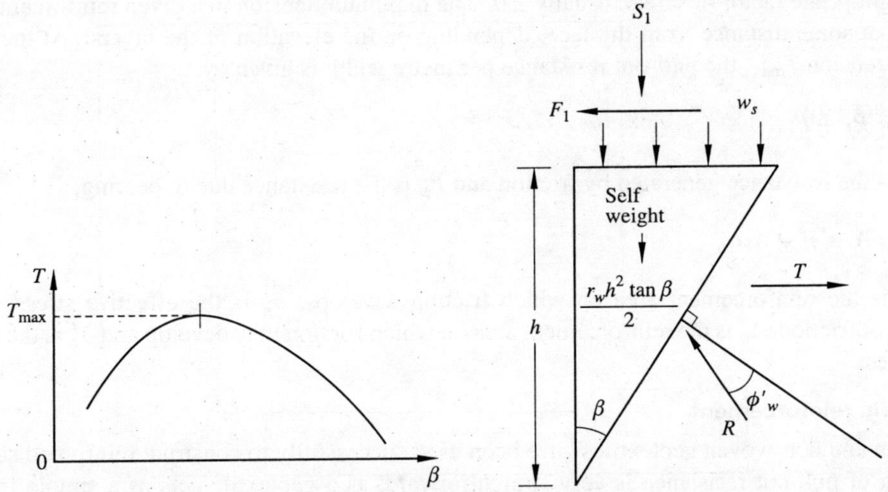

Fig. 6.12 *Variation of tensile force with wedge angle*

Mobilizing force

The above assumptions help in estimating the force that mobilizes the pull-out as well as the total tension T that is necessary to keep the wedge in equilibrium. In a general case the weight of the wedge W is

$$T = \frac{W}{\tan(\phi'_w + \beta)} \tag{6.11}$$

In the simple case of a vertical wall with surcharge (i.e., $F_1 = 0, S_1 = 0$) constructed with cohesionless soil from equilibrium of forces, the total force (T) is given by

$$T = \frac{h \tan(\gamma_w h + 2w_s)}{2\tan(\phi'_w + \beta)}. \tag{6.12}$$

The maximum value of T occurs corresponding to the angle of inclination of the wedge given by $\beta = 45° - \frac{\phi'_w}{2}$.

For a general case,

$$T = \left[\frac{W \tan\beta(\gamma_w h + 2w_s) + 2S_1}{2\tan(\phi'_w + \beta)}\right] + F_1. \tag{6.13}$$

Resisting force

Pull-out resistance of the reinforcement needs to be high enough, both locally and globally. Local checks compare the maximum tension with the maximum available bond force in each individual layer. Global checks consider potential failure mechanisms developing in the reinforced zone and compare the overall driving and resisting forces. Pull-out resistance should be more than the above force T with an appropriate factor of safety, usually 2.0. The maximum tension in a given reinforcement layer will occur at some distance from the face, depending on the elevation of the layers. At the point of maximum tension $T_{\max}$, the pull-out resistance per metre width is given by

$$P_p = P_f + P_b, \tag{6.14}$$

where P_f is the resistance generated by friction and P_p is the resistance due to bearing.

$$P_p = A_f \sigma'_n \mu + A_b \sigma'_b \tag{6.15}$$

where A_f is the reinforcement area on which friction develops, σ'_n is the effective stress, μ is the coefficient of friction, A_b is the reinforcement area on which friction may develop and σ'_b is the effective bearing stress.

Sheet/fabric reinforcement

Both woven and non-woven geotextiles have been used successfully to construct reinforced soil walls. Calculation of pull-out resistance is very straightforward, as a geotextile acts as a simple full-width rough sheet:

$$P_p = 2\int_0^{L_a} \sigma'_n f_b \tan \phi'_w dL. \tag{6.16}$$

Hence the pull-out resistance per metre width for a layer of geotextile at depth z below a level ground surface with uniform surcharge is:

$$P_p = 2L_a(\gamma z + w_s) f_b \tan \phi'_w. \tag{6.17}$$

In practice the bond coefficient f_b is determined from direct shear or pull-out tests on reinforced soil; values in the range of 0.7-1.0 have been measured. Geotextile manufacturers' literature should be consulted for product-specific values.

Polymer-grid reinforcement

Polymer grids (geogrids) generally have a smaller aperture size than wire meshes, and interact with soil in a highly efficient (though rather more complex) manner. Jewell and Wroth (1984) have demonstrated that friction and bearing components may be satisfactorily combined into a single bond coefficient. The formulations with regard to the role of bond have been discussed in Chapter 5 (Eq. (5.4)). The bond coefficient f_b is

$$f_b = a_s \frac{\tan \delta}{\tan \phi'_w} + a_b \frac{B}{2S} \frac{N_q}{\tan \phi'_w} \left[\frac{\sigma'_b}{\sigma'_n}\right], \tag{6.18}$$

where a_S is the fraction of grid surface area that is solid,
a_b is the fraction of grid width available for bearing,
δ is the angle of skin friction,
B is the diameter/thickness of transverse members,
S is the spacing of transverse members,
N_q is the bearing capacity factor.

Jewell (1996) indicates that for S/a_bB ratios in the range 10 to 20, geogrids behave as perfectly rough sheets with f_b values close to 1.0 (typical calculations are given in Chapter 5). The pullout resistance (per metre) for a geogrid at depth z below a level ground surface with uniform surcharge is defined in the same way as was done for a geotextile earlier.

Consider the situations shown in Fig. 6.13 As indicated earlier, the pull-out resistance available in a geogrid placed at a depth h_i resists the mobilising force T. For a level backfill, it is determined by

$$T = \frac{L_{ip} 2 f_b \tan \phi'_w (\gamma_w h_i + 2 w_s)}{FS}. \tag{6.19}$$

The check should be performed considering the surcharge and without the surcharge as well. A factor of safety of 2 is to be used.

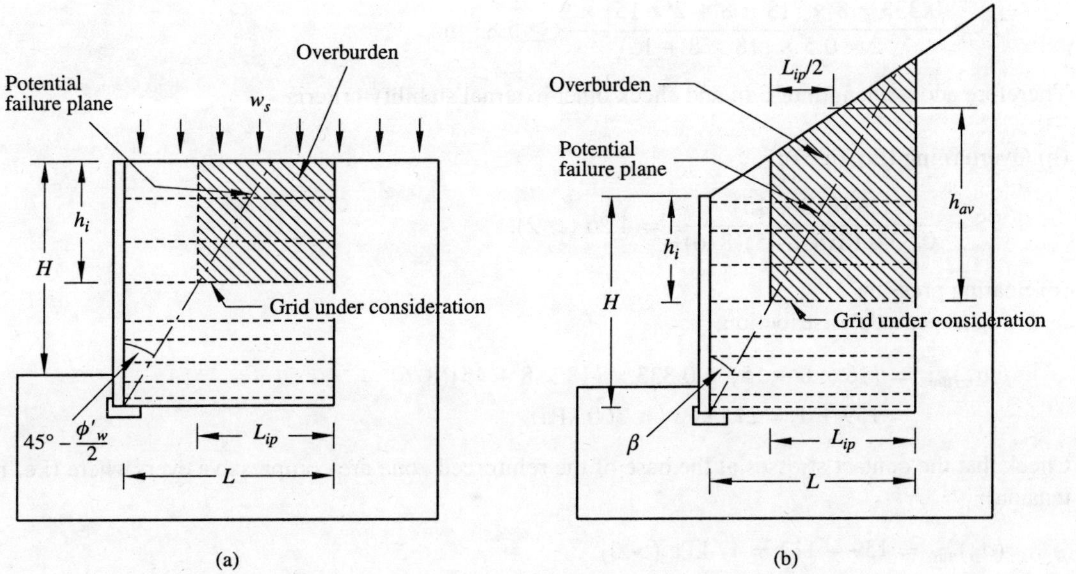

Fig. 6.13 *Consideration of pull-out resistance for (a) level backfill and (b) sloping backfill*

For a sloping fill (no surcharge)

$$T = \frac{L_{ip} 2 f_b \tan \phi'_w (\gamma_w h_{av})}{FS}. \tag{6.20}$$

For a sloping fill, the overburden corresponding to a height h_{av} as shown in Fig. 6.13 should be considered.

Bar reinforcement (soil nails)

Similar to strips, normal stresses act around the circumference of a bar (πd) rather than just on the upper and the lower surfaces; pull-out resistance for a single bar is calculated taking this into consideration.

Example 6.1 An 8 m high wall is to be built using sand fill and polymer-grid reinforcement. The sand has $\phi' = 30°$, $\gamma = 18$ kN/m^3 and is to be used for the wall and the backfill. A surcharge loading of 15 kPa is to be allowed for; the maximum safe bearing pressure for the foundation soil is 300 kPa. Three grids of different design strengths are available: grid A at 24 kN/m, grid B at 16.5 kN/m and grid C at 10.5 kN m (all have a bond coefficient f_b of 0.9). The fill will be compacted in layers 250 mm thick.

External stability

(a) Outward sliding (Eqns (6.1) and (6.2))

$$K_a = (1 - \sin 30°)/(1 + \sin 30°) = 0.333$$

$$\mu = f_b \tan\phi = 0.9 \times \tan 30° \approx 0.5.$$

For a factor of safety of 2.0 against sliding, the minimum length of layers is:

$$L \geq \frac{0.333 \times 8 \times (18 \times 8 + 2 \times 15) \times 2}{2 \times 0.5 \times (18 \times 8 + 15)} \geq 5.83 \text{ m}.$$

Therefore adopt a length of 6 m and check other external stability criteria.

(b) Overturning (Eq. (6.3)):

$$\text{FS} = \frac{3(18x8 + 15)}{0.333(18x8 + 45)(8/6)^2} = 4.26\ (> 2).$$

(c) Bearing pressure

Using trapezoidal distribution,

$$(\sigma_v)_{\max} = (18 \times 8 + 15) + 0.333 \times (18 \times 8 + 45)(8/6)^2$$
$$= 159 + 1 = 271 \text{ kPa } (< 300 \text{ kPa}).$$

Check that the contact stresses at the base of the reinforced zone are compressive everywhere (i.e. no tension):

$$(\sigma_v)_{\min} = 159 - 112 = 47 \text{ kPa } (> 0)$$

Internal stability

Combining Eqs. (6.6) and (6.7) and substituting the values, the maximum tension in a layer at depth z below the top of the wall is given by

$$T_1 = 0.333\left[(18z + 15) + 0.333(18z + 45)(z/6)^2\right] S_V$$

or

$$(S_V)_{\max} = \frac{P_d}{0.333\left[(18z + 15) + 0.333(18z + 45)(z/6)^2\right]}$$

For three different grids that are available, calculations using the above equation have been carried out. The results are plotted in the form of spacing versus depth as shown in Fig. E6.1. It is given that the compaction thickness is 250 mm and hence the spacing needs to be in multiples of compaction thickness. The results in Fig. E6.1 show that grid A with higher strength can be used up to 6 m at a spacing of 25 cm, beyond which the lesser strength geogrids B and C can be used to ensure that the tension in the geogrid does not exceed the design strength.

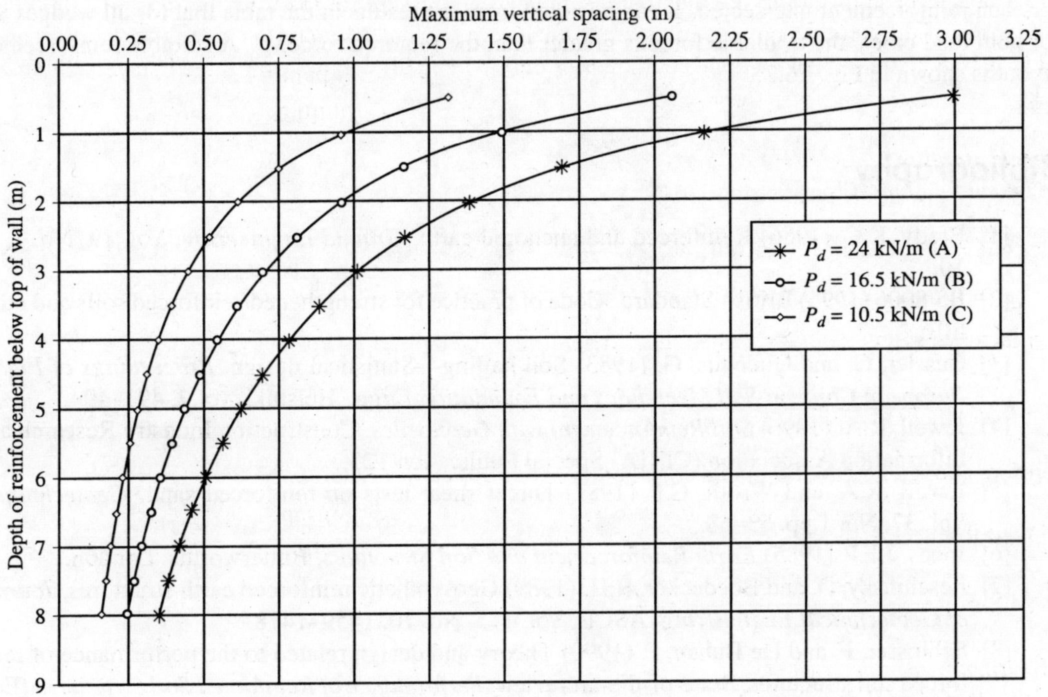

Fig. E6.1 *Spacing versus depth relationship for grids A and B.*

Wedge stability check

Select trial wedges at depths 1 to 8 m below the top of the wall and calculate the total required force T. Carry out check with and without surcharge w_s. For a critical wedge angle $\beta = (45° - \phi'_w/2) = 30°$ for a wedge of height h (Eq. (6.12)), the total tension force T is given by

$$T = \frac{h \tan 30°(18h + 2 \times 15)}{2 \tan(30° + 30°)} = 3h^2 + 5h.$$

Without surcharge, $T = 3h^2$.

For a reinforcing layer at depth z below the top of the wall, the pull-out resistance is given by

$$P_P = 2[L - (h - z) \tan \beta] \times (\gamma z + w_s) \times 0.9 \times \tan 30°/2.$$

The factor 2 in the numerator denotes the upper and lower surfaces on either side of the geogrid and factor 2 in the denominator refers to the factor of safety.

$$P_P = 2[6 - (h - z)\tan 30°] \times (18z + 15) \times 0.9 \times \tan 30°/2.$$

The above calculations are shown in the following Table E6.2. From pull-out failure considerations, the available force is taken as the lesser of the pull-out resistance P_P and the design tensile strength P_d for each reinforcement intersected. It can be noted from the results in the table that for all wedges and for both load cases, the available force is greater than the required force, T. A suitable reinforcement layout is shown in Fig. E6.2.

Bibliography

[1] Brady, K.C. (1986) Reinforced and anchored earth, *Ground Engineering*, Vol. 19, No. 7, 7–10.

[2] BS 8006 (1995) British Standard: Code of practice for strengthened/reinforced soils and other fills.

[3] Gassler, G. and Gudehus, G. (1983) Soil nailing—Statistical design, *Proceedings of Eighth European Conf. on Soil Mechanics and Foundation Engg.*, Helsinki, vol.2, 491–494.

[4] Jewell, R.A. (1996) *Soil Reinforcement with Geotextiles*, Construction Industry Research and Information Association (CIRIA) Special Publication 123.

[5] Jewell, R.A. and Wroth, C.P. (1987) Direct shear tests on reinforced sand, *Geotechnique*, Vol. 37, No. 1 pp. 53–68.

[6] Jones, J.F.P. (1985) *Earth Reinforcement and Soil Structures*, Butterworths, London.

[7] Leschinsky, D. and Boedecker, R.H. (1989) Geosynthetic reinforced earth structures, *Journal of Geotechnical Engineering*, ASCE, Vol. 115, No. 10, 1459–1478.

[8] Schlosser, F. and De Buhan, P. (1990) Theory and design related to the performance of reinforced soil structures, State-of-the-art review, *Performance of Reinforced Soil Structures* (Eds. A. McGown, K.C. Yeo and K.Z. Andrewes), Glasgow, Thomas Telford, 1–14.

Table E6.2

Wedge depth (m)	Force to be resisted T (kN/m)		Type of grids $A = 24$ kN m $B = 16.5$ kN m $C = 10.5$ kN m	Position of grids from top of wall, z (m)	Effective length (L_e)	Design tensile strength, P_d (kN/m)	Sum of pull-out resistance, P_p (kN/m)		Total available force (kN/m) lesser of P_d & P_p	
	With surcharge	Without surcharge					With surcharge	Without surcharge	With surcharge	Without surcharge
2				0.25	4.99	10.5	50.56	11.67		
			$2C + 1B$	1	5.42	10.5	92.98	50.72		
				1.75	5.86	16.5	14.49	95.84		
	22	12		Total		37.5	285.03	158.23	37.5	37.5
4				0.25	3.83	10.5	38.86	8.97		
				1	4.27	10.5	73.18	39.92		
				1.75	4.70	16.5	113.59	76.94		
			$2C + 4B$	2.5	5.13	16.5	160.06	120.05		
				3	5.42	16.5	194.42	152.16		
				3.5	5.42	16.5	194.42	152.16		
	68	48		Total		87	811.59	585.00	87	87
6				0.25	2.68	10.5	27.16	6.27		
				1	3.11	10.5	53.38	29.12		
				1.75	3.55	16.5	85.69	58.04		
				2.5	3.98	16.5	124.06	93.05		
				3	4.27	16.5	153.02	119.76		
			$2C + 5B + 3A$	3.5	4.27	16.5	153.02	119.76		
				4	4.85	16.5	219.04	181.27		
				4.5	5.13	24	256.10	216.08		
				5	5.42	24	295.86	253.59		
				5.5	5.71	24	338.32	293.80		
	138	108		Total		175.5	1737.30	1400.15	175.5	175.5
8				0.25	1.53	10.5	15.46	3.57		
				1	1.96	10.5	33.58	18.32		
				1.75	2.39	16.5	57.79	39.14		
				2.5	2.82	16.5	88.06	66.05		
				3	3.11	16.5	111.62	87.36		
				3.5	3.40	16.5	137.88	111.36		
				4	3.69	16.5	166.84	138.07		
				4.5	3.98	24	198.50	167.48		
				5	4.27	24	323.86	199.59		
			$2C + 5B + 11A$	5.5	4.27	24	323.86	199.59		
				6	4.85	24	309.68	271.19		
				6.25	4.99	24	330.57	291.68		
				6.5	5.13	24	352.14	312.12		
				6.75	5.28	24	374.38	333.24		
				7	5.42	24	397.29	355.03		
				7.25	5.57	24	420.89	377.50		
				7.5	5.71	24	445.15	400.64		
				7.75	5.86	24	470.10	424.64		
	232	192		Total		367.5	4412.69	3831.91	367.5	367.5

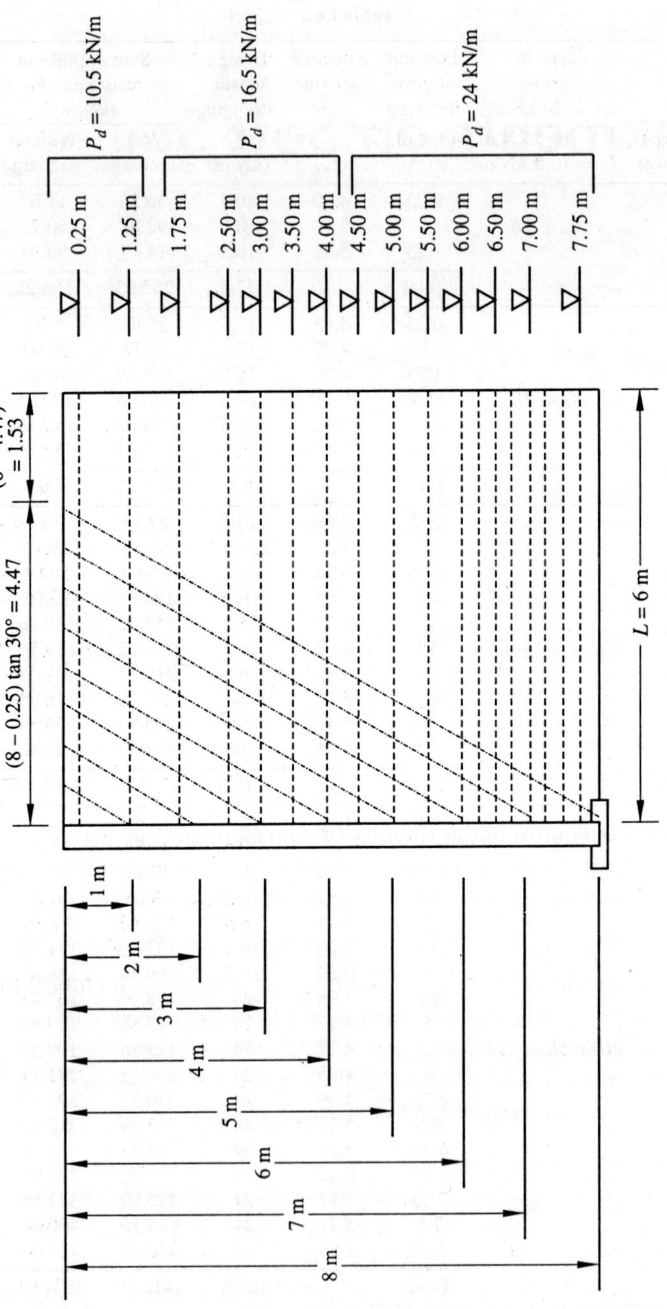

Fig. E6.2 *Reinforcement layout*

7 Embankments on Soft Soil

7.1 Introduction

Soft soils usually occur in estuarine or marshy areas. They exhibit considerable volume changes upon loading, and are associated with poor shear strength. If the foundation soil for an embankment is too soft to support both the dead load and live loads on the embankment, some type of treatment on it is necessary. Methods such as excavation and replacement, stabilisation with preloading and the use of prefabricated wide drains have been used traditionally. However, in recent years the use of reinforcement techniques to improve embankment stability, has become very popular owing to a number of successful applications of this technique. Holtz (1994) indicated that reinforcement at the basal level in the form of geotextiles, steel or precast concrete blocks or automobile tire casings/chipped tires significantly improves the embankment stability, and this aspect was demonstrated by a number of case studies carried out in Sweden and Denmark for both highway and railway embankments. The use of geosynthetics to improve the performance of embankments on soft soil is a very effective form of reinforced soil technique. BS 8006 (1995) divides the applications (Fig. 7.1) into two categories.

1. Use of reinforcement to control initial stability of embankment (Fig. 7.1(a)). The reinforcement is used as (a) basal reinforcement beneath embankment, (b) basal reinforcement with vertical drains, and (c) basal mattress reinforcement. Some applications can be found in Humphrey and Holtz (1986).

2. Use of reinforcement to control both initial stability and settlement of embankment(Fig. 7.1(b)). Nowadays applications are growing in number owing to the requirement of construction of high-speed railway embankments on soft soil, where deformation requirements are critical. Gartung and Verspohl (1996) report a case study of a geogrid-reinforced railway embankment on piles constructed on soft soil close to Berlin in Germany.

This chapter focuses on the use of reinforcement to control initial stability. Figure 7.2 shows the corresponding ultimate limit states and serviceability limit states that need to be considered for design of basal reinforced embankments. Ultimate limit states corresponding to local stability (Fig. 7.2(a)) to overall stability (Fig. 7.2(e)) need to be addressed for both unreinforced embankments and reinforced embankments. Figures 7.2(g) and 7.2(f) show the serviceability limit states. In particular, when the use of geogrids/geotextiles is considered, the roles of other ultimate limit states such as rotational stability, lateral spreading, foundation extrusion and serviceability limits assume importance. Reinforcement using geotextiles/geogrids as basal reinforcement is useful in the following ways (Jewell, 1996):

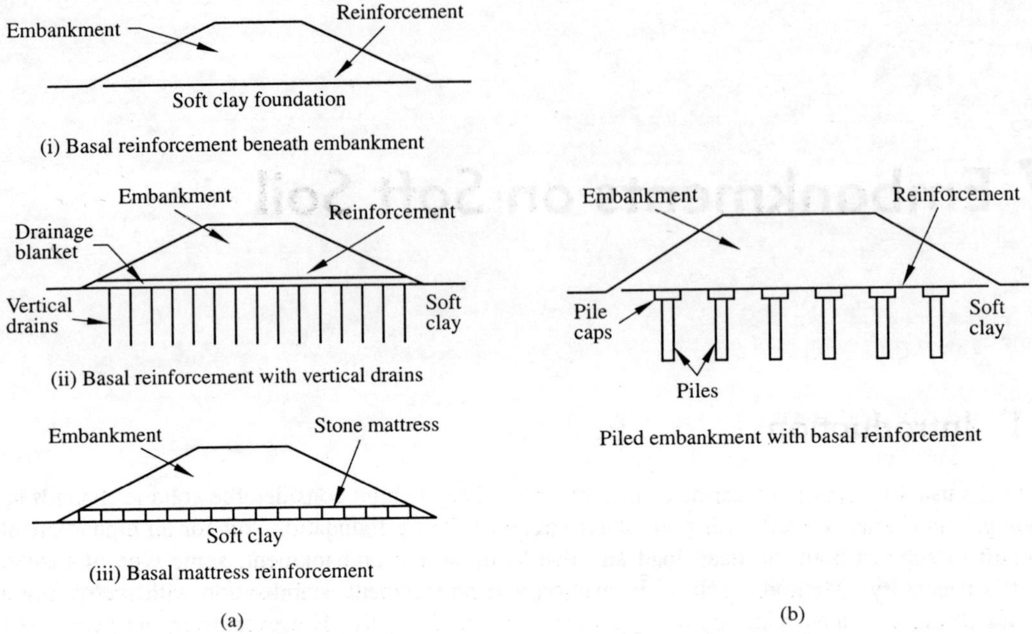

Fig. 7.1 *(a) Use of reinforcement to control the initial stability of the embankment: (i) basal reinforcement beneath embankment, (ii) basal reinforcement with vertical drains, and (iii) basal matterss reinforcement. (b) Use of reinforcement to control both initial stability and settlement of embankment (BS 8006, 1995)*

1. It improves embankment stability.
2. It permits controlled construction over soft soils.
3. It ensures more uniform settlement of embankment.

Reinforcement is very effective when placed at or close to the foundation surface. If the reinforcement were absent, the factor of safety at the end of construction falls below unity, as shown in Fig. 7.3.

7.2 Analysis

For design purposes, Jewell (1996) considers the modes of failure in the following manner:

(i) Internal stability: If the reinforcement is at the interface of soft soil and embankment material, direct sliding is likely to occur at the interface (Fig. 7.2(c)).

(ii) Overall stability: Overall stability is likely to be affected due to (a) slippage across the foundation surface, and (b) stability in the foundation (Fig. 7.2(d)).

For design purposes, information concerning the variations of shear strength with depth is necessary. Two idealised distributions are possible, depending on the nature of soil profile. (a) Uniform variation of undrained shear strength with depth, and (b) increase of shear strength with depth. The first distribution is relevant if the soft soil is of limited depth and the second distribution is applicable in cases

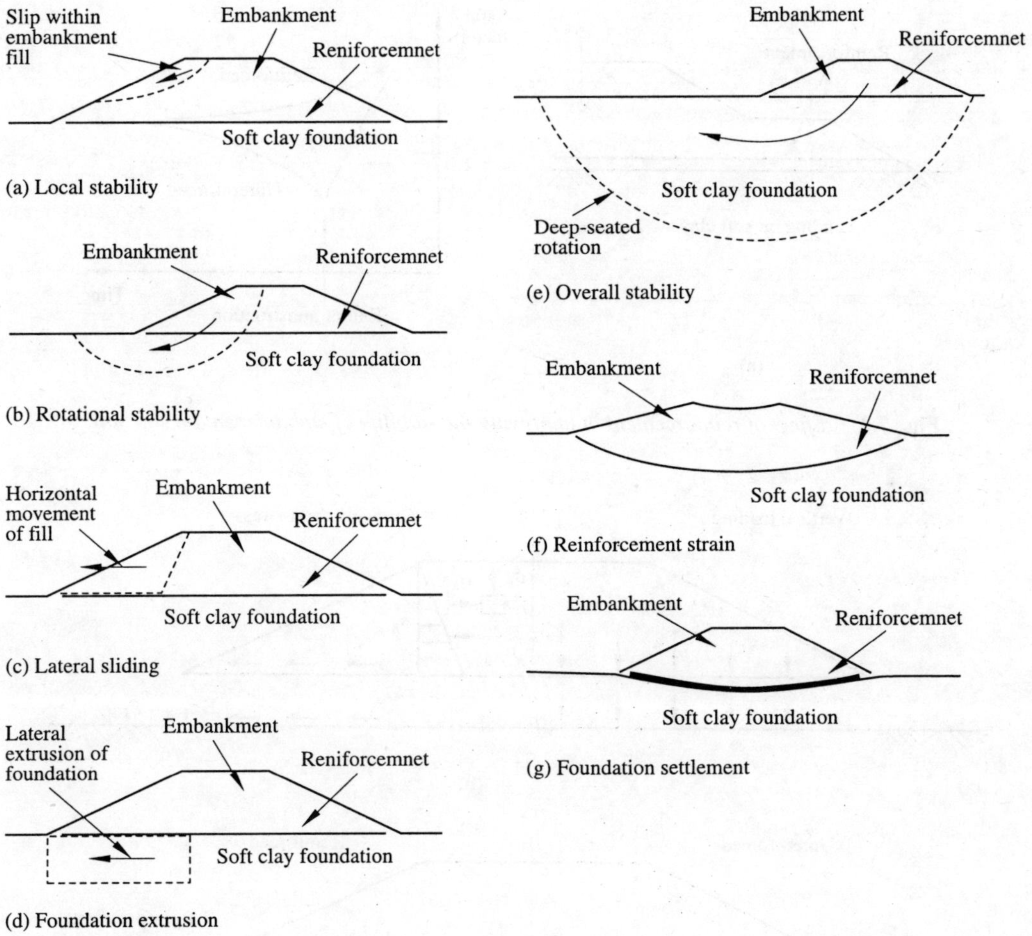

Fig. 7.2 *Ultimate limit states and serviceability limit states that need to be considered for design of basal reinforced embankments (BS 8006, 1995)*

where soft soil extends to larger depths compared to the height of the embankment. Two types of loads are considered for equilibrium as shown in Fig. 7.4.

1. Vertical loading (due to self-weight of the soft soil and external loading)
2. Outward directed lateral force P_{fill}

To prevent the lateral spreading of the embankment, horizontal stresses within the fill must be balanced by shear reactions at the base. In an unreinforced embankment, the lateral thrust P_{fill} is transferred to the foundation, but when the reinforcement is present, some of the thrust is carried by the reinforcement. If the reinforcement is sufficiently strong and stiff, the reinforcement could carry the entire lateral thrust, as shown in Fig. 7.4.

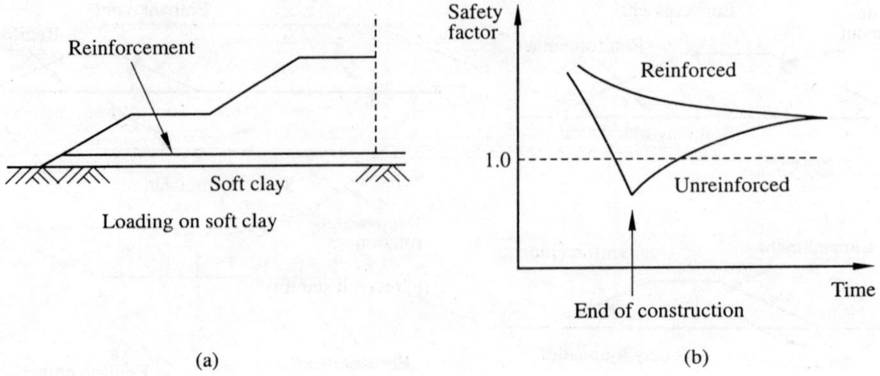

Fig. 7.3 *Effect of reinforcement in improving the stability of embankment on soft soil*

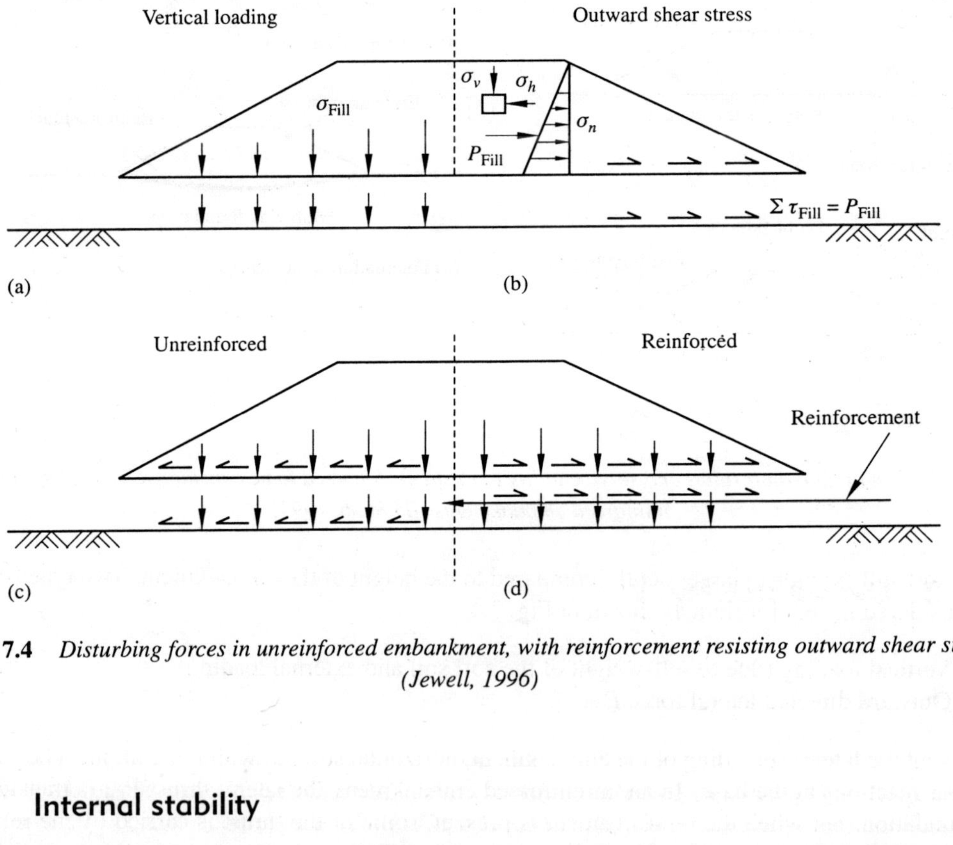

Fig. 7.4 *Disturbing forces in unreinforced embankment, with reinforcement resisting outward shear stress (Jewell, 1996)*

7.2.1 Internal stability

Inclusion of reinforcement in soil can allow preferential direct sliding to occur along the surface of the reinforcement layer (Fig. 7.5). Slippage results if the available resistance at the interface is not sufficient to support the lateral thrust from the fill. If the design parameters for the embankment

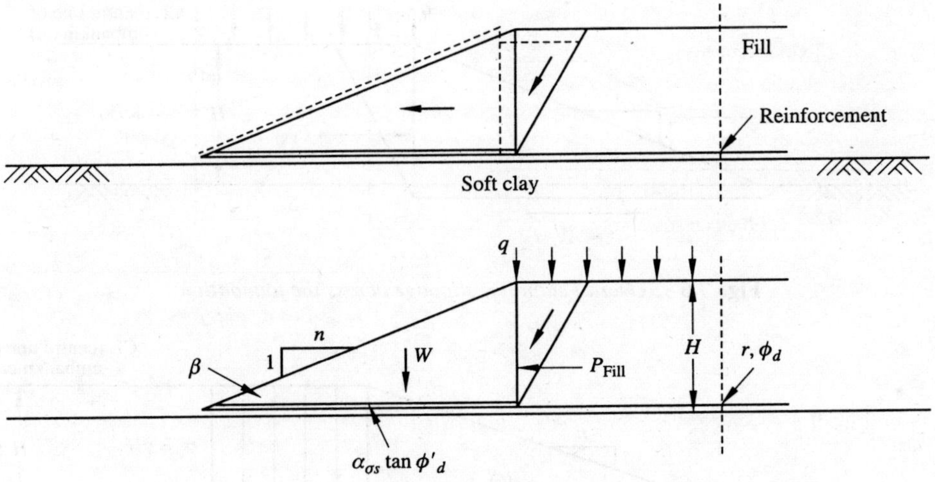

Fig. 7.5 *Internal stability: direct sliding failure in the fill (Jewell, 1996)*

material are (i) friction angle = ϕ_d, (ii) unit weight, height and slope of embankment = γ, H, $(1:n)$ respectively, (iii) Surcharge = q, (iv) coefficient of direct sliding = α_{ds}, then the relationships for direct sliding resistance at the interface and active thrust from the fill are as follows.

Direct sliding resistance at the interface = $\alpha_{ds} \tan\phi_d \times \gamma n H^2/2$.

Active thrust from the fill = $P_{\text{fill}} = K_{ad}(\gamma H^2/2 + qH)$.

For equilibrium,

$$\alpha_{ds} \tan\phi_d \times \gamma n H^2/2 = K_{ad}(\gamma H^2/2 + qH),$$

$$n > \frac{K_{ad}}{\alpha_{ds}\tan\phi_d}\left(1 + \frac{2q}{\gamma H}\right). \tag{7.1}$$

7.2.2 Overall stability

Overall stability needs to be considered in terms of

- slippage across the foundation surface (Fig. 7.6),
- stability in the foundation (Fig. 7.7).

Slippage sliding or slippage could occur across the surface of soft clay in an unreinforced zone close to reinforcement location; it could also occur if the reinforcement member breaks. From equilibrium considerations, shear resistance in the plane of slippage in the soil is equal to the lateral thrust form the fill. Hence,

$$S_{ud} n H = K_{ad}(\gamma H^2/2 + qH),$$

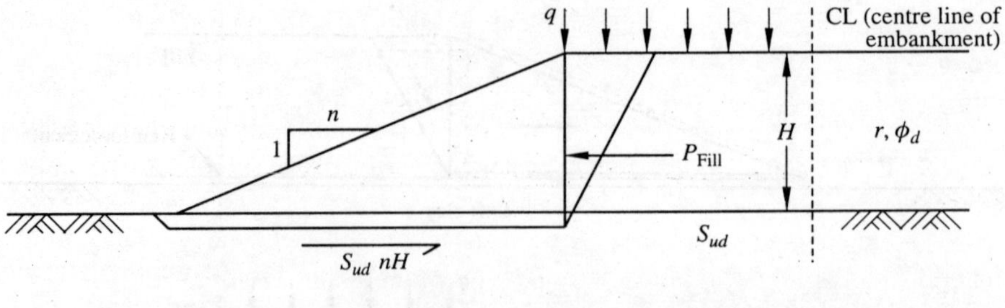

Fig. 7.6 *Overall stability; slippage across the foundation*

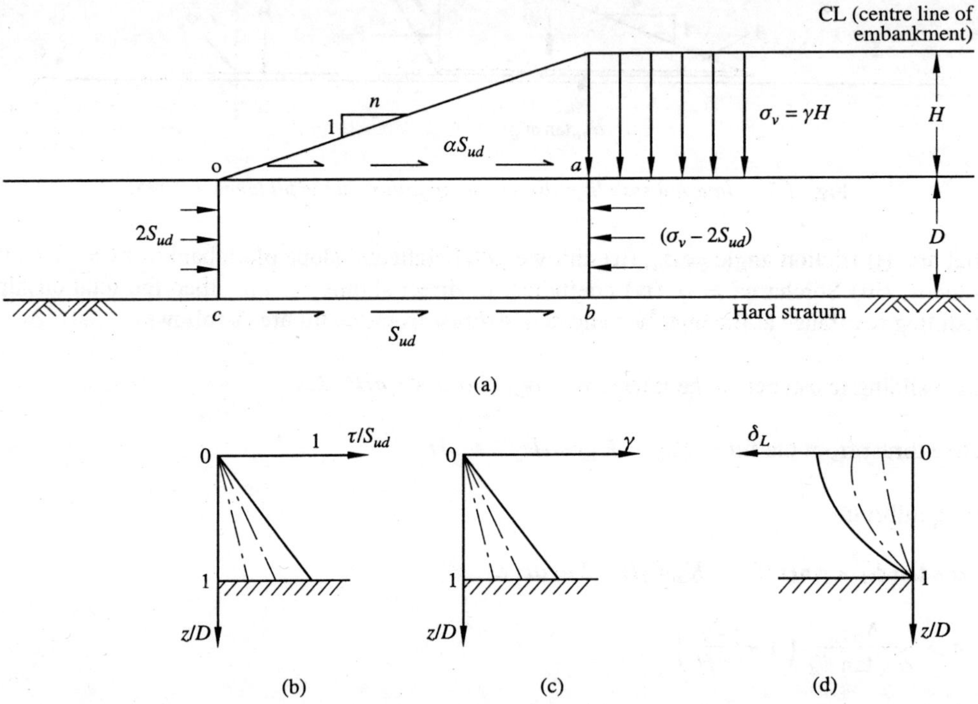

Fig. 7.7 *Foundation stability: (a) applied stresses, (b) shear stresses (c) shear strain, (d) shear displacement for $\alpha = 0$ (Jewell, 1996)*

$$n > \frac{K_{ad}}{2S_{ud}}\gamma H\left(1 + \frac{2q}{\gamma H}\right). \tag{7.2}$$

7.2.3 Stability in the foundation

Vertical embankment loading causes an increase in the vertical stress in the foundation soil and a corresponding increase in the horizontal stress. The lateral stress developed eventually causes the foundation soil to get displaced laterally beneath the embankment side slope. The response is schematically shown in Fig. 7.7.

Consider a zone of soft clay beneath the embankment side slope. Since the soft soil is of limited depth D, it may be assumed that the vertical stress $\sigma_v = \gamma H$ acts uniformly and that the principal axes of stress on either side of the block *oabc* are horizontal and vertical. The soil block *oabc* experiences a net horizontal stress $(\sigma_v - 2S_{ud})$ on *ab* and $2S_{ud}$ on *oc*.

For the case where there are no outward or inward shear stresses acting on the tip of soil block (*oa*), $\alpha = 0$. The lateral sliding of the soil block is resisted by the available shear strength on the horizontal plane, $(S_{ud} \times nH)$. Hence,

$$nH.S_{ud} = (\sigma_v - 2S_{ud})D - 2S_{ud}.$$

$$\text{Hence } n = \frac{\gamma D}{S_{ud}} - \frac{4D}{H}. \tag{7.3}$$

Equation (7.3) assumes that the reinforcement supports outward shear stresses from the lateral fill such that the net lateral shear stress across the foundation is zero.

7.3 Influence of Reinforcement Extensibility

Equation (7.3) is derived considering that the net shear stress at the foundation level is zero ($\alpha = 0$). Mobilization of shear stress at the interface is an important factor in the efficiency of the reinforced embankments. The shear stress could mobilize inward, outward or become nil, depending on the extensibility of the reinforcement, shear strength of the soil and loading conditions. Three possibilities are denoted as follows:

(a) $\alpha = -1$, full outward shear stress
(b) $\alpha = 0$, zero shear stress
(c) $\alpha = 1$, full inward shear stress

The case of zero shear stress $\alpha = 0$ at the foundation level is valid if the net extension in the reinforcement caused by the axial tension mobilised to support the fill is compatible with the lateral deflection at the surface of the soft clay. The magnitude of lateral displacement at the clay surface and the net lateral displacement gives an idea of the effectiveness of the reinforcement. If the reinforcement is totally effective, full inward shear stresses develop and the net lateral displacement in the clay surface is zero and is characterized by $\alpha = 1$. On the other hand, if there is greater extension in the reinforcement than the lateral deflection in the soft clay, outward shear stresses are mobilised on the clay surface and at the extreme (full outward shear stresses are characterized by $\alpha = -1$) and the lateral deflection is maximum. The above concept and the understanding of the role of extensibility, proposed by Jewell (1996), is illustrated in Fig. 7.8. It may be noted that, in practice, some shear strain mobilization is necessary for the development of reinforcement force in the member and hence $\alpha = 1$ is not practical; lesser values are likely to represent real situations in the field, indicating the effectiveness of reinforced embankments.

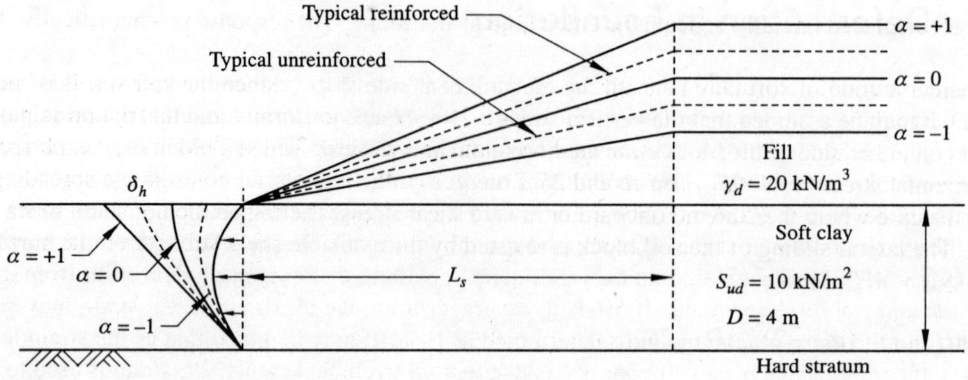

Fig. 7.8 *Effect of reinforcement extensibility on foundation stability and deformation (Jewell, 1996)*

7.4 Relationships for Design

The following relationship holds good for the design of an embankment built on a foundation of uniform undrained shear strength S_{ud} and limited depth D (Jewell 1996).

Embankment height

$$H = \frac{4D}{\left\{\frac{\gamma D}{S_{ud}} - (1+\alpha)\,n\right\}}. \tag{7.4}$$

Factor of safety in a reinforced embankment

$$\mathrm{FS} = \frac{S_{ud}}{\gamma H}\left(4 + (1+\alpha)\,\frac{nH}{D}\right). \tag{7.5}$$

Lateral deflection at the ground surface

$$\delta_h = \frac{S_{ud}D}{2G}(1-\alpha). \tag{7.6}$$

The maximum reinforcement to provide stability is

$$P_R = \gamma H^2\left(\frac{\alpha n D}{4D + (1+\alpha)\,nH} + \frac{K_{ad}}{2}\right), \tag{7.7}$$

where

$$K_{ad} = \frac{1-\sin\phi_d}{1+\sin\phi_d}.$$

In the case of unreinforced embankment,

$$\mathrm{FS} = \frac{S_u}{\gamma H}\left(\frac{8D + 2nH}{2D + K_{ad}H}\right). \tag{7.8}$$

7.5 Deformation in Foundation

A small amount of deformation across the interface of geosynthetic and embankment soil is necessary to mobilize the tensile force in the reinforcement. However, once the interfacial friction is mobilised between the soil and reinforcement, inward shear stresses develop which prevent the lateral spreading of the embankment material. The modulus of the geosynthetic material controls the spreading. The shear strain developed at the interface initiates lateral spreading; the distribution of shear strain varies, with the maximum value at the interface and the minimum at a deeper level in the soil. Fowler and Haliburton (1980) indicated that an average lateral spreading of 5% is reasonable both from the construction and stability viewpoint. If 5% is the average strain, the maximum shear strain may be taken as 10% and the geosynthetic modulus determined at 10%. It may be noted that as the strain level decreases, the modulus increases. If cohesive soils are used in embankments, 2% strain is used to reduce the possibility of tensile cracking of soil. The load induces lateral deformation in the foundation soil which is also dependent on the extensibility of the reinforcement.

Maximum lateral displacement $= \delta_h = \gamma_{\max} D/2 = \tau_{\max} D/2G$, where $\gamma_{\max}$ = maximum shear strain and $\tau_{\max}$ = maximum shear stress = S_u = Shear strength of the soil and G is the shear modulus of the soil.

7.5.1 Settlement analysis

Conventional settlement analysis methods for embankments on soft soils are followed. The creep of reinforced embankments in the soft soil should be considered in terms of the rate of creep of geosynthetic versus the rate of consolidation and strength gain of the foundation. If the foundation soil consolidates and gains strength at a faster rate than the rate at which the geosynthetic loses its strength, then there is no problem. In any case, each of these issues need be thoroughly examined considering the actual soil properties and geosynthetic characteristics, as well as the temperature and other boundary conditions such as drainage, presence of a thin layer of sand in the foundation soil, etc.

7.6 Overall Stability with respect to Bearing

To assess the overall stability of the embankment, it is necessary to verify that the ultimate bearing capacity of the soft soil (q_{ult}) is more than the applied loading intensity (Jones, 1996). For a soft soil the ultimate bearing capacity is given by

$$q_{\text{ult}} = 5.14 c_u.$$

At the base of the embankment, the loading intensity is

$$\frac{W}{2B'} + w_s',$$

where W is the weight of the embankment, $2B'$ is the total width of the embankment and w_s' is the surcharge pressure at the base, deduced using (1 : 2) distribution.

$$\text{Factor of safety} = \frac{5.14 c_u}{(W/2B' + w_s')}. \tag{7.9}$$

Geocells have been used as basal reinforcements in many situations to improve embankment stability (Fig. 7.9). Jones (1996) presents the design procedure for a geocell reinforced embankment assuming that a rigid foundation mattress at the interface of embankment and soft soil alters the normal slip circle plane by forcing it through the mattress. As a result, the failure surface is deepened and shifts to the stiff underlying layer present below the soft soil of limited depth. Krishnaswamy et al. (2000) conducted model studies on geocell embankments in the laboratory. The influence of various parameters such as tensile stiffness of geogrids used for fabricating the geocell material, the height and size of the geocell layer, length of the geocell layer, and the type of fill material inside the geocell, on the behaviour of the embankments was investigated through a series of laboratory tests.

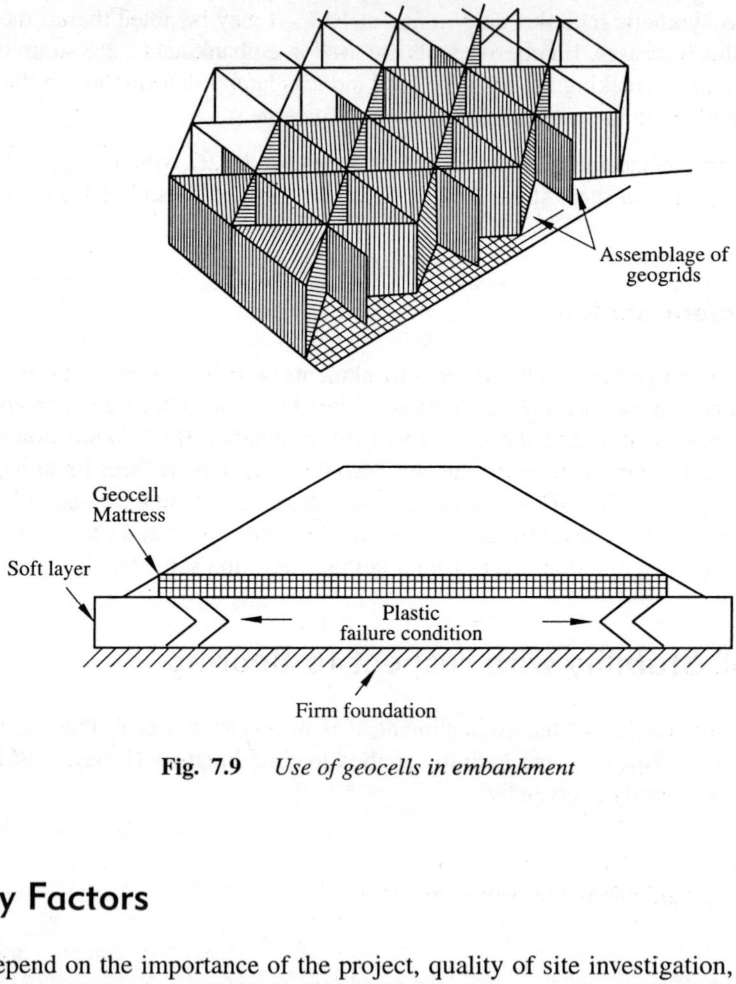

Fig. 7.9 *Use of geocells in embankment*

7.7 Safety Factors

Safety factors depend on the importance of the project, quality of site investigation, variation of soil as well as geosynthetic properties. While partial factors are necessary for soil as well as geosynthetic materials, an overall factor of safety of 1.3 is considered acceptable to prevent failure. With regard to loss of strength of the geosynthetic due to creep, values of 40% to 60% of the ultimate stress are considered satisfactory. This information should be obtained from isochronous curves of the geosynthetic material.

Example 7.1 A 200 m long and 4 m high road embankment with a crest width of 15 m needs to be constructed over soft clay of 4 m depth ($S_{ud} = 15$ kN/m^2, constant with depth). The embankment is constructed rapidly in one stage. The end-of-construction stability needs to be ensured. Granular fill in the nearby area ($\phi_{cv} = 33°$, $\gamma_d = 20$ kN/m^3) is used. An overall FS = 1.3 may be used.

Geosynthetics with index strength in the range of 200 to 400 kN/m are available. Index strength is to be corrected for loading period (equal to 90% of consolidation) and for the temperature corresponding to ground conditions ($t = 20°$C). Under these conditions the manufacturers' data shows that corresponding to the end of 90% consolidation (time = 2.5 years), the reference strength is 60% of the design strength. An allowance for compaction and damage can be given in the form of $f_d = 1.2$, besides an allowance of $f_{env} = 1.1$ for environmental degradation.

Arrive at the suitable reinforcement properties. Make the necessary stability checks.

Solution Factor of safety of reinforced embankment is given by

$$FS = \frac{S_{ud}}{\gamma H}\left\{[4 + (1+\alpha)]\frac{nH}{D}\right\}.$$

One can use $H = 4.5$ m, though the design height is 4 m. This gives an additional factor of safety and also takes care of profile requirements.

$$1.3 = \frac{15}{20 \times 4.5}\left\{[4 + (1+\alpha)]\frac{n \times 4.5}{4}\right\}$$

$$\Rightarrow 7.8 = 4 + (1+\alpha)1.125n \Rightarrow 3.38 = (1+\alpha)n.$$

For $\alpha = 0, \quad n = 3.38$

$\alpha = 0.5, \quad n = 2.25$

$\alpha = 1.0, \quad n = 1.69$

Unreinforced case FS

$$FS = \frac{S_{ud}}{\gamma H}\left(\frac{8D + 2nH}{2D + K_{ad}H}\right).$$

$$= \frac{15}{20 \times 4.5}\left(\frac{8 \times 4 + 2n \times 4.5}{2 \times 4 + 0.294 \times 4.5}\right).$$

Corresponding to the two values of $\alpha = 0$ and 0.5, the n values are 3.38 and 2.25 respectively. The factor of safety without reinforcement is given by the above formula. Both the values are less than the targeted value of 1.3 and hence use of reinforcement is necessary.

For n	FS	
2.25	0.94	< 1.3.
3.38	1.12	

The reinforcement force is given by

$$P_R = \gamma H^2\left(\frac{\alpha n D}{4D + (1+\alpha)nH} + \frac{K_{ad}}{2}\right)$$

Substitution of the value of above quantities leads to $P_R = 59.54$ kN/m for $\alpha = 0.0$ and 117.97 kN/m for $\alpha = 0.5$.

As indicated in Section 7.5, in order to mobilise design shearing resistance S_{ud}, the shearing strain γ (or γ_{max}) in the range of 10% for soft clay should be considered. It may also be noted that it is possible to arrive at appropriate shear strain level if the stress–strain characteristics of the soft soil and the reinforcement are known considering the concept of strain compatibility presented in Chapter 2. Hence,

Required stiffness $= J_{req} = P_R/\gamma_{max}$, and $\gamma_{max} = 10\%$

For $\alpha = 0.0$, $J_{req} = 595$ kN/m and for $\alpha = 0.5$, $J_{req} = 1180$ kN/m

The shear modulus in given by $G = 1/\gamma$ and using

$$\delta_h = \frac{S_{ud}D}{2G}(1 - \alpha),$$

we get,

$$\delta_h = 200 \text{ mm for } \alpha = 0.0$$
$$\delta_h = 100 \text{ mm for } \alpha = 0.5$$

showing that the stiffer/higher reinforcement force will reduce that lateral displacement.

It is given that geosynthetics are available with index strengths in the range of 200 to 400 kN/m and that the index strength is to be corrected for the loading period (equal to 90% of consolidation) and for the temperature corresponding to ground conditions ($t = 20°$C). Under these conditions, manufacturer's data shows that at the end of 90% consolidation (time = 2.5 years), the reference strength is 60% of the design strength. An allowance for compaction and damage, $f_d = 1.2$, besides an allowance of $f_{env} = 1.1$ for environmental degradation is to be made.

The following calculations are made to examine the suitability of the available geosynthetic materials.

For index strength of $P_{index} = 200$ kN/m :

Reference strength $= 0.6 \times 200 = 120$

Field strength $= 120/1.2 \times 1.1 = 91$

Design strength $= 91/1.3 = 70$

Available stiffness $= 70/0.10 = 700$ kN m

The required stiffness values for $\alpha = 0$ and $\alpha = 0.5$ are 595 kN/m and 1180 kN/m respectively. If the displacement is not critical (if $\alpha = 0$ is acceptable), the material with 200 kN m serves the purpose. If the displacement needs to be controlled, it can be noted that 400 kN/m is acceptable as the available stiffness is 1400 kN/m and is more than the required value of 1180 kN/m.

For a soft soil, the ultimate bearing capacity is given by $q_{ult} = 5.14\, c_u = 5.14 \times 15 = 77.1$ kN/m^2. At the base of the embankment the loading intensity for $\alpha = 0.5$ is

$$\frac{W}{2B'} + w_s = \frac{(2.25 \times 4 \times 4 + 4 \times 15)20}{33} + 0 = 58.18 \text{ kN/m}^2.$$

The loading intensity is less than the ultimate bearing capacity and hence safe.

Bibliography

[1] BS 8006 (1995) Code of Practice for strengthened/reinforced soils and other fills.

[2] Fowler, J. and Haliburton, T.A. (1980) Design and construction and analysis of fabric reinforced embankments, *The use of geotextiles for soil improvement*, ASCE Convention, Portland, 89–118.

[3] Gartung, E. and Verspohl, J. (1996) Geogrid reinforced railway embankment on piles monitoring, *Earth Reinforcement Conf.*, Kyushu (Eds. H. Ochiai et. al) 209–214.

[4] Humphrey, D.N. and Holtz, R.D. (1986) Reinforcement—review of case histories, *Geotextiles and Geomembranes*, vol. 4, No. 2, 129–144.

[5] Holtz, R.D. (1994) Design and construction of embankments on very soft soils, *Geosynthetic World* (Ed. J.N. Mandal), Wiley Eastern Ltd.

[6] Jewell, R.A. (1996) *Soil Reinforcement with Geotextiles*, CIRIA publication.

[7] Jones, C.J.F.P. (1996) *Earth Reinforcement and Soil Structures*, Thomas Telford Publishing, London.

[8] Krishnaswamy N.R., Rajagopal K., Latha G.M. (2000) Model studies on geocell supported embankments constructed over a soft clay foundation, *Geotechnical Testing Journal*, ASTM, Vol. 23, No. 1, 45–54.

8 Soil Nailing

8.1 Introduction

The fundamental concept of soil nailing consists of reinforcing the ground by closely spaced passive inclusions to create *in situ* a coherent gravity structure, and thereby increase the overall shear strength of the *in situ* soil and restrain its displacements. The basic design consists of transferring the resisting tensile forces generated in the inclusions into the ground through friction mobilized at the interfaces. The technique allows the engineer to efficiently use the *in situ* ground to provide vertical support, and presents significant advantages over conventional rigid retaining walls or external bracings. The first soil-nailed wall using modern soil nailing techniques was built near Versailles in France in 1972. The technique included installing high-density, grouted soil nails into a 18 m high wall and facing it with reinforced concrete. Since then, this technology has improved to the point where soil nail construction is fast becoming the preferred method of building soil retaining walls. In particular, France and Germany, continues to lead the world in soil nail technology. Use of soil nail construction is increasing in popularity in the United States and the Federal Highway Administration (FHWA) has been implementing this technology on highway projects such as road widening since 1980s. This technique is now being used in many parts of the world very successfully. Considerable research work using fullscale experiments with instrumentation was undertaken in France and Germany to establish reliable analytical methods and design codes for soil nailing. In France, a national project CLOUTERRE ('clou' means nail, and 'terre' means soil) was conducted and the research findings were published (CLOUTERRE 1991). Good reviews of the technique are available from Bruce and Jewell (1987), Elias and Juran (1991), Abramson (1994) and Schaefer (1997). A number of applications of this technique have been possible. Fig. 8.1 shows some applications of this technique. These include:

- Stabilization of railroad and highway cut slopes
- Excavation of retaining structures in urban areas for high-rise buildings and underground facilities
- Tunnel portals in steep and unstable stratified slopes
- Construction and retrofitting of bridge abutments with complex boundaries involving wall support under piled foundations.

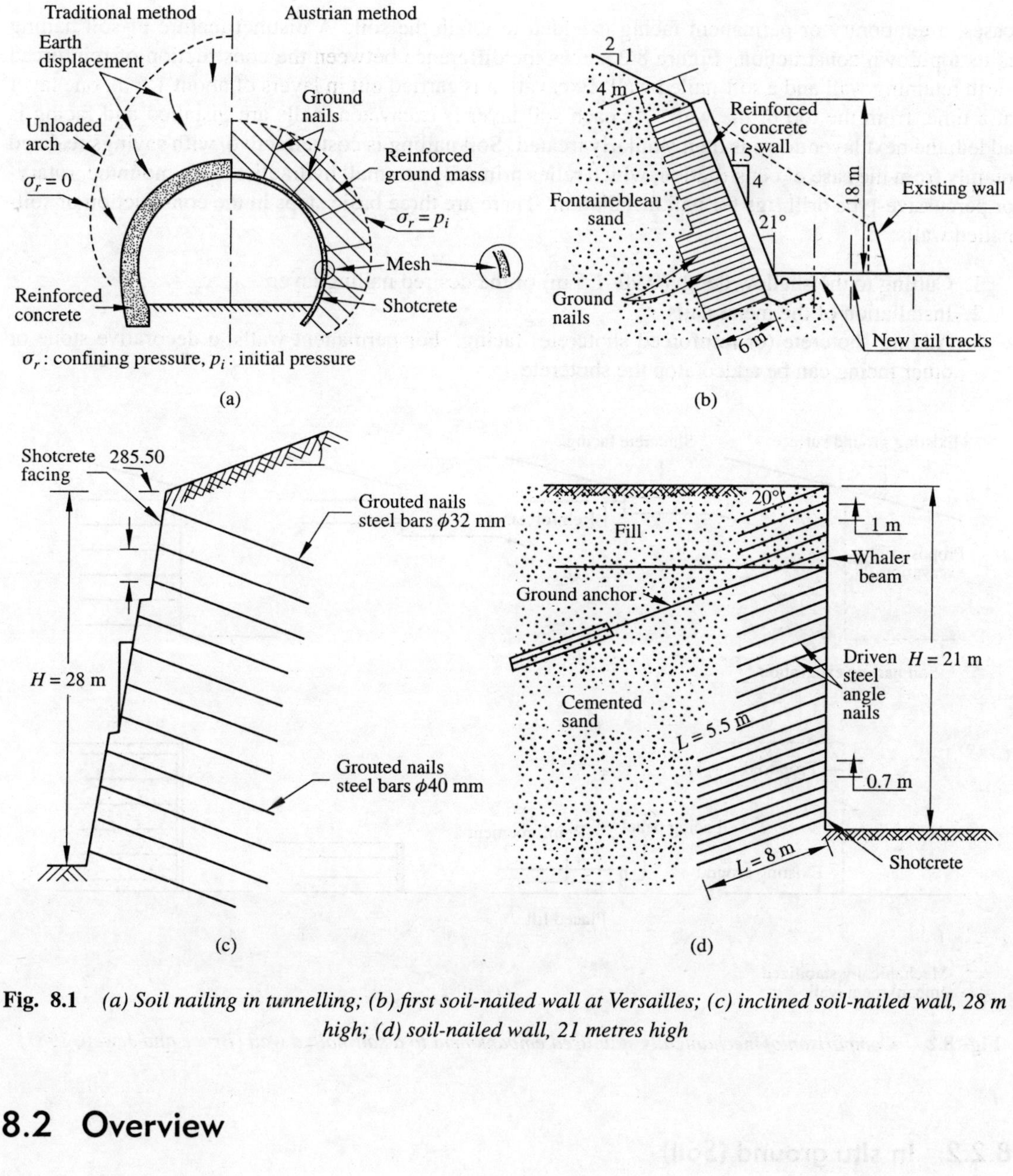

Fig. 8.1 *(a) Soil nailing in tunnelling; (b) first soil-nailed wall at Versailles; (c) inclined soil-nailed wall, 28 m high; (d) soil-nailed wall, 21 metres high*

8.2 Overview

8.2.1 Technique

Soil nailing is a method of construction which reinforces the existing ground. Inclusions in the form of nails are inserted into the soil in a closely-spaced pattern to increase its overall shear strength. The nails develop tension as the ground deforms laterally in response to the ongoing excavation. In most cases, a temporary or permanent facing is added to retain the soil. A distinct feature of soil nailing is its top-down construction. Figure 8.2 shows the difference between the construction of reinforced earth retaining wall and a soil-nailed wall. Excavation is carried out in layers of about 1.5 m, one layer at a time, from the top of the wall. As each soil layer is excavated, nails are installed and facing is added; the next layer down is then similarly treated. Soil nailing is cost-effective, with savings realized mainly from the ease of construction, which relies primarily on small hydraulic, track-mounted, rotary- or percussive-type drill rigs for nail installation. There are three basic steps in the construction of soil-nailed walls:

1. Cutting to the shallow depth (0.90–1.5 m) of the desired nailing layer
2. Installation of the metal nails
3. Adding shotcrete (or reinforced shotcrete) facing. For permanent walls, a decorative stone or other facing can be added atop the shotcrete

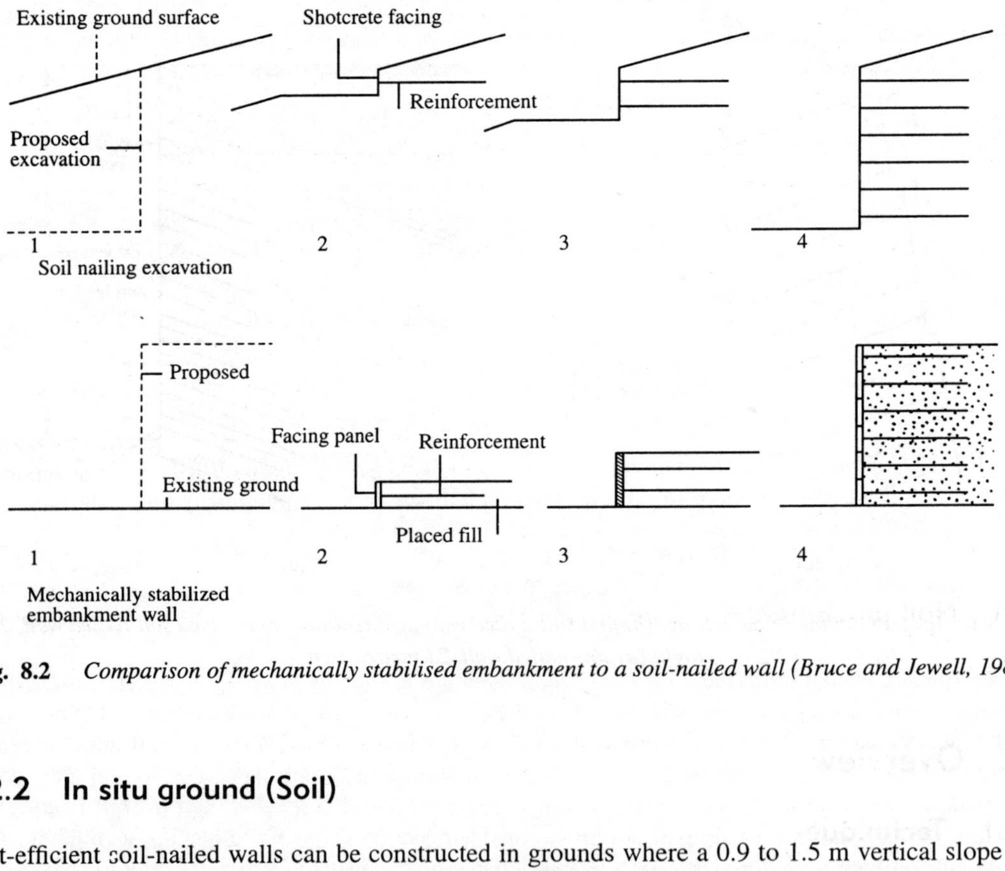

Fig. 8.2 *Comparison of mechanically stabilised embankment to a soil-nailed wall (Bruce and Jewell, 1987)*

8.2.2 In situ ground (Soil)

Cost-efficient soil-nailed walls can be constructed in grounds where a 0.9 to 1.5 m vertical slope can stand without support for up to two days during construction and is stable for the few hours it takes to drill and insert the nails. The depth of the cut layer depends on the soil's ability to stand unsupported while the nails are being inserted. Weathered rock, silts, clays with low plasticity that are not prone to creep, naturally cemented sands and gravels, heterogeneous and stratified soils, and fine to medium homogeneous sands are suitable for soil nail construction. Soils not conducive to soil nail technology are soft plastic clays, peat/organic soils, loose, low-density, and/or saturated soils; and coarse sand and gravels that are uncemented. Hence soil analysis is essential prior to soil nail construction. In the analysis, it is necessary to check if the soil can lead to corrosion of the nails so that proper care in the form of coating or provision of sacrificial thickness can be examined. Corrosion prevention is necessary in permanent structures and in 'aggressive' soils, which are defined as having a pH below 4.5, a resistivity below 2,000 ohm-cm, sulfate levels above 200 ppm, and chloride levels above 100 ppm. If these conditions exist, corrosion-protected nails must be used.

8.2.3 Nails

Soil nails are installed in a pattern so as to ensure both internal and external stability of the wall. The nails used in construction are generally steel bars that resist tensile and shear stresses and bending moments. Most projects are designed to use nails with a uniform length and cross-sectional area. Nail length is usually about 60–80% of the height of the wall, depending on soil conditions. Prior to construction, it is useful to conduct tests on nails to determine nail–soil adhesion and their resistance to pull-out failure. Several types of soil nails are currently in use.

Driven nails are generally small-diameter nails (15–46 mm) with a relatively limited length (to about 20 m), made of mild steel that are closely spaced in the wall (two to four nails per square meter). Nails with an axial channel can be used to permit the addition of grout sealing. Driven nails are the quickest (four to six per hour) and most economical to install with a pneumatic or hydraulic hammer.

Jet-grouted nails are a composite of grouted soil and a central steel rod, up to 40 cm thick. Nails are installed using a high-frequency vibro-percussion hammer, and cement grouting is injected during installation. This method has been shown to increase the pull-out resistance of the composite, and the nails are corrosion-resistant. Cement grout with a water-to-cement ratio of about 0.4 to 0.5 is usually used.

Launched nails (25 and 38 mm in diameter and up to 6 m or longer) are fired directly into the soil with a compressed-air launcher. Used primarily for slope stabilization, this technique involves the least site disturbance. More information on this is available from Schaefer (1997).

8.2.4 Nail placement

Typically, nails of equal length and cross-sectional area are uniformly spaced. In general, for drilled and grouted nails, spacing is one nail per 0.9 to 2 m, both vertically and horizontally. Driven nails require higher densities of as much as one-and-a-half to two nails per square meter. Nail rows are often staggered to increase face stability. The angle of inclination is generally between 10 and 20°. Nail length depends on several factors, including soil strength, soil–nail adhesion, and overall loading of the system. In general, minimum nail length is considered to be about 0.6 times the wall height for vertical walls with no backslope. Shorter nails have been used in walls in stiff soils. The experience in Europe indicates that it might be preferable in some cases to install longer and higher-capacity nails in the upper two-thirds of the wall, as research shows that this reduces wall displacement.

8.2.5 Facing

Once the nails are installed and grouted, a shotcrete facing between 75 mm to 150 mm thick is applied, with a wire mesh at mid-thickness. Facing is not considered to be a structurally significant supporting part of the wall. The experience in France indicates that nail loads at the facing generally do not exceed 30–40% of the maximum loads in the nail, so they recommend a facing designed for a uniform wall pressure equal to 60% of the maximum nail load with a nail spacing of 0.9 m. For walls with greater nail spacing (e.g., 3 m), the facing should be designed for 100% of the maximum nail load. Permanent structures can be made attractive and aesthetically pleasing, with the addition of cast-in-place concrete facings with a minimum of 200 mm thickness. Precast decorative panels may also be attached directly to the shotcrete facing.

8.2.6 Drainage

Drainage is a critical element in planning and construction. Most commonly, face drainage is used and a drainage element is placed behind the shotcrete wall covering the nailed structure. The drainage elements are installed from top to down as construction proceeds. Typically, synthetic strips or perforated pipes can be installed with appropriate spacing. The water is collected at the wall base and channeled away. Alternatively, weep holes can be made through the face of the wall, used with or without perforated drainpipes. Whichever method is used, it is necessary to channel the water away from the wall so that it does not collect behind it.

8.3 Advantages and Disadvantages of Soil Nails

Soil-nailed walls compare favourably in terms of costs, speed and ease of construction with other soil-retention systems such as sheet piles, tie back walls, etc. Tieback walls, for example, require structural facing elements that are pre-tensioned and anchored to the ground with steel that is strong and stiff enough to hold the soil structure behind it. The anchors must be tensioned with enough load to support the facing without creep or other failure. Unlike tiebacks, soil-nailed walls are not tensioned; they are passive reinforcements in that that once placed within the soil of the wall, they create a coherent gravity mass. Thus, soil nails create a condition of internal stability within the wall; stability does not depend on the strength of the outer facing but is generated within the structure itself. Further, soil nail construction eliminates the need for placing timber lagging or sheet piling, as well as the need for costly facing systems. The nail length is shorter than that used in tiebacks. In mechanically stabilized earth (MSE) walls, which are somewhat akin to soil-nailed walls, the greatest amount of stress accumulates at the bottom of the wall through compaction. Thus, the lower parts of an MSE wall are most likely to deform or fail. In a soil-nailed wall, the greatest stress is initially contained within the upper layers of the wall, it then passes downward as construction proceeds. Unlike MSE walls, however, the placement of more and/or longer nails in the upper portion of the soil nail wall seems to limit the transfer of tension to the lower wall.

Soil nail construction has many advantages:

- Soil-nailed walls can be built to follow curved or zigzagged outlines.
- The equipment used is easily portable and can fit easily into small spaces.

- The process is flexible and makes modifications easy to carry out (e.g., spacing of nails can be altered as needed during construction).
- Construction causes less noise and traffic obstruction on highways.
- The process creates less impact on adjacent or nearby properties than do other construction methods.
- It generally requires less space and manpower.

Soil nail construction also has its drawbacks. They are:

- The method cannot be used at sites where groundwater is a problem.
- It is inappropriate for sites with soils having very low shear strength, in sand and gravels that lack cohesion, and on sites with other unsuitable soils.
- Soil must be able to stand unsupported while it is being nailed and before shotcrete application.
- Good drainage is essential, especially for permanent structures and in places prone to freeze–thaw cycles.
- It is imperative that projects be designed, constructed, and monitored by specialists with experience in soil-nailed wall construction.

8.4 Soil–Nail Interaction

In soil nailing, the load transfer mechanism and the ultimate pull-out resistance of the nails depend primarily upon the soil type and its strength characteristics, installation technique, drilling method, size and shape of the drilled hole, as well as grouting method and pressure used. Estimates of the pull-out resistance of nails are based upon empirical formulae (or ultimate interface shear stress values) derived from field experience. These formulae are useful for feasibility evaluation and preliminary design. Table 8.1 (Elias and Juran 1991) provides a summary of estimated ultimate interface shear stress values for soil nails as a function of soil type and installation technique.

8.5 Engineering Behaviour

Several full-scale experiments carried out (CLOUTERRE 1991, Gassler 1993) showed that the distribution of tension along the nails in a soil-nailed wall was fairly similar to that observed and calculated with the finite element method in reinforced earth walls. Generally, the shape and position of the line of maximum tension, which can be considered as a potential failure surface, are very different from the straight line assumed from Coulomb theory of earth pressures. In the case of a wall with a vertical facing and a soil with a horizontal surface (Fig. 8.3), this line is situated nearer to the facing and is curved in shape, with, at its head, a vertical part at an approximate distance of $0.3H$ from the facing (H being the height of the wall). However, the inclination of the facing with respect to the vertical and the presence of a slope at the head of the wall will modify the position of the line of maximum tension.

With regard to the state of earth pressure, the variation of the state of stress in the soil of a soil-nailed wall, along with maximum tension $T_{\max}$, is studied by considering the ratio K defined by

$$K = \frac{T_{\max} \cos \theta}{\gamma z S_v S_h},$$

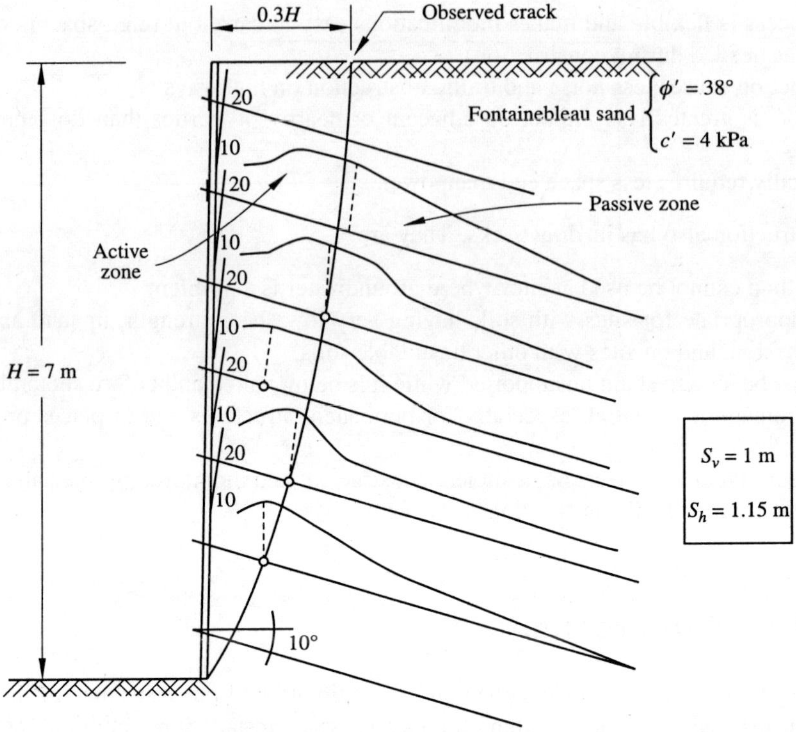

Fig. 8.3 *Locus of maximum tension in the reinforcement soil (CLOUTERRE, 1991)*

where, γz is the overburden pressure above the point of maximum tension, S_v and S_h are vertical and horizontal spacings between nails and θ the inclination of the nails with respect to the horizontal. Experiments conducted on real structures, small-scale models, and finite elements calculations show that the state of stress is near K_0 (coefficient of lateral earth pressure at rest) at the head of a soil-nailed wall, while at the bottom it is lower than the coefficient of active earth pressure K_a (Fig. 8.4).

Further, the studies indicate that bending stiffness of the nails does not have significant effect and the surcharge does not alter the locus of the maximum tension line (Gassler 1993). The research also indicated that in spite of the differences in inclusions, methods of installation and soil profiles, maximum facing displacement does not exceed 0.3% of the structure height (Schaefer 1997).

8.6 Design Procedure

The design procedure for a soil-nailed retaining structure should include the following steps:

1. For the specified structural geometry (depth and cut slope inclination), ground profile, and boundary (surcharge) loadings, estimate the working nail forces and location of the potential sliding surface.

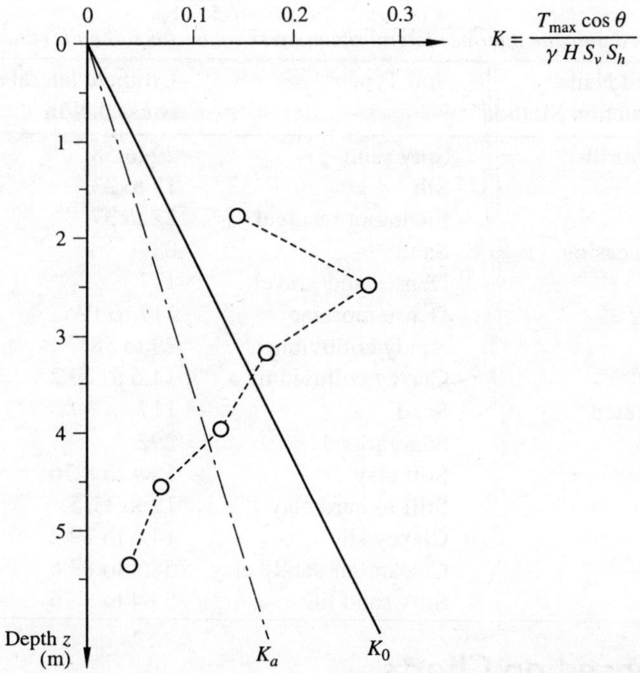

Fig. 8.4 *Sate of stress in a reinforced soil-nailed wall (CLOUTERRE, 1991)*

2. Select the reinforcement type (cross-sectional area, length, inclination, and spacing) and verify local stability at each reinforcement level, that is, verify that nail resistance (strength and pull-out capacity) is sufficient to withstand the estimated working forces with an acceptable factor of safety.
3. Verify that the global stability of the nailed-soil structure and the surrounding ground is maintained during and after excavation with an acceptable factor of safety.
4. Estimate the system of forces acting on the facing (i.e., lateral earth pressure and nail forces at the connection) and design the facing for specified architectural and durability criteria.
5. For permanent structures, select corrosion protection relevant to the site conditions.
6. Select the drainage system for groundwater piezometric levels.

The available design methods for soil-nailed retaining structures are based on limit equilibrium methods or modified slope stability analyses. These are used to evaluate the global safety factor of the nailed structures with respect to a rotational or translational failure along potential sliding surfaces, taking into account the shearing, tension, or pull-out resistance of the inclusions crossing the potential failure surface. Similar to traditional slope stability analysis, limit equilibrium conditions are used to search for the most critical failure surface, which can be located either inside or outside the soil-nailed retaining structure. The available design procedures involve different assumptions with regard to the shape of the failure surface, the type of soil–reinforcement interaction, and the resisting force in the inclusions. In the following sections, some details on some popular methods are presented.

Table 8.1 *Estimated ultimate interface lateral shear stress values for soil nails (Elias and Juran, 1991)*

Grouted Nails Construction Method	Soil Type	Ultimate lateral shear stress, kN/m
Rotary drilled	Silty sand	29 to 58
	Silt	17 to 23
	Piedmont residual	22 to 37
Driven casing	Sand	88
	Dense sand/gravel	117
	Dense moraine	117 to 175
	Sandy colluvium	29 to 58
	Clayey colluvium	14.6 to 29.2
Jet grouted	Sand	117
	Sand/gravel	292
Augured	Soft clay	5.84 to 8.76
	Stiff to hard clay	12 to 17.5
	Clayey silt	14.6 to 29.2
	Calcareous sandy clay	58.4 to 87.6
	Silty sand fill	5.84 to 8.76

8.7 Design based on Charts

8.7.1 Gassler method

Stocker et al. (1979) developed a design method (also termed as the German Method, as it has been developed and is widely used in Germany), assuming a bilinear sliding surface. Gassler and Gudehus (1981) presented some results and applications. In this method, the shear resistance of the soil as defined by Mohr–Coulomb criterion, is assumed to be entirely mobilised along the potential failure surface. Only reinforcement tension forces are considered. The available resisting nail forces and the total required force to maintain equilibrium are considered. The available resisting forces are the forces beyond the assumed failure surface. The total force required to maintain limit equilibrium is obtained by considering the polygon of forces acting on a rigid soil wedge/failure surface limited by the potential failure surface. The resisting forces are provided by the pull-out capacity of the nails. Gassler (1996) suggests that for nailed walls and slopes in both the reinforced and reinforced zones, plastic shear deformations are located in thin shear planes, whereas the greater part of the soil is rigid in the limit condition and hence the consideration of rigid body mechanics is satisfactory. With regard to critical failure mechanisms, Gassler (1996) suggests that in soils with little cohesion, both slip circle and two-part wedge mechanisms are equivalent, and for soils with moderate cohesion, slip circle is the critical failure mode. Gassler developed design charts that cover different inclinations of slope (α), different angles corresponding to end of nails (δ), design load (q_d) and specific nailing density μ, given by

$$\mu = \frac{T}{\gamma S_v S_h},$$

where γ is the unit weight ($\mathrm{kN/m^3}$), S_v and S_h are the vertical and horizontal spacings and T is the factored (estimated or measured) pull-out force. A typical chart is shown in Fig. 8.5. Gassler (1996) also advocates the use of partial safety factors format as per the Eurocode, wherein the loads such as driving forces and surcharges are increased and resistances in the form of soil friction angle are reduced by suitable factors. It may be noted that for preliminary designs, design charts present an expedient solution which can be examined later by other methods as well. The use of the design charts of Gassler is illustrated by the following example.

Example 8.1 Consider a 10 m high, near-vertical slope in an urban setting in a sandy soil. The ground water table is very deep, and beyond the bottom of excavation. The following parameters are given. The unit weight of the soil is 18 $\mathrm{kN/m^3}$, surcharge pressure is 23 $\mathrm{kN/m^2}$ and the friction angle is 36.25°. It is proposed to use 20 mm diameter steel bars of 6 m length for soil nailing. The site requirements are that the line through which the nails end needs to be vertical and the vertical spacing shall be 1.0 m. The wall will have an inclination of 10°.

Solution An appropriate design chart corresponding to the geometric parameters of the nailed wall, such as $\ell/h = 0.6$, inclination $(\alpha) = 10°$ and the line through lower nail ends vertical ($\rho = 0$) is chosen as indicated (left column, second row of Fig. 8.5).

The concept of partial factor of safety is followed in this example. In this approach, loads such as surcharge pressures are increased and resisting forces (factors such as friction angle) are decreased by appropriate factors. A partial factor of safety of 1.25 is chosen for the friction angle. Hence the design friction angle is

$$\phi_d = \frac{36.25}{1.25} = 29°$$

The surcharge stress is given as 23 $\mathrm{kN/m^2}$. This is increased by multiplying by a factor of 1.3. Hence, normalised design load is

$$= \frac{q_d}{\gamma h} = \frac{23 \times 1.3}{18 \times 10} = 0.17.$$

Corresponding to 0.17, one obtains a specific nailing density of $\mu_d = 0.87$ from the chart as indicated in Fig. 8.5. The specific nailing density μ_d is multiplied by 1.20 as per the partial safety factor approach. Hence characteristic specific nailing density is

$$0.87 \times 1.25 = 1.09$$

The pull-out resistance of the nail can be estimated by assuming that the pull-out failure is represented by Mohr–Coulomb failure criterion. Corresponding to the last nail located at 9.5 m depth, the pull-out resistance is given by

$$T = (\gamma h_i)\pi dL \tan\phi_{\mu}.$$

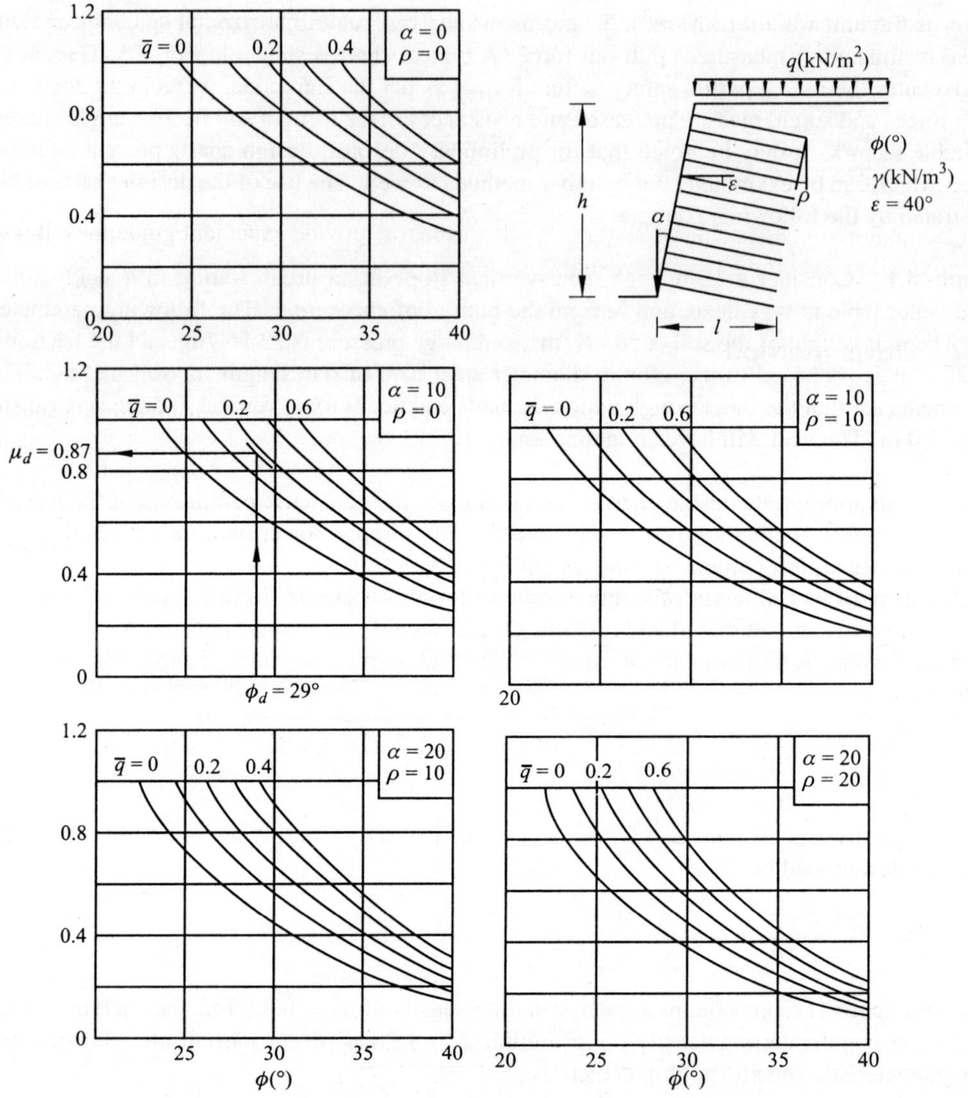

Fig. 8.5 *Design charts for nailed walls and slopes in non-cohesive soil with nail length $l/h = 0.60$ and with constant surcharge $q^+ = (q/\gamma h)$ acting on horizontal crest (inclination of slope (α), inclination of nail end with vertical (ρ), q surcharge) (Gassler, 1996)*

Consider $\phi_\mu = \phi_d$. Hence,

$$T = 18 \times 9.5 \times \pi \times 0.02 \times 6 \times \tan 29^\circ = 35.73 \text{ kN}.$$

Using a partial safety factor of 1.3, characteristic pull-out resistance

$$T_c = \frac{35.43}{1.3} = 27.48 \text{ kN}.$$

The vertical spacing is 1.0 m. Hence, the corresponding horizontal spacing can be obtained from

$$S_h = \frac{T_c}{\gamma_d \times S_v \times \mu_c} \qquad \left[\mu_c = \frac{T_c}{\gamma_d S_h S_v}\right]$$

$$= \frac{27.48}{18 \times 1 \times 1.08} = 1.40 \text{ m}$$

Hence adopt horizontal spacing of 1.40 m. While the charts provide expedient guidelines, it is always necessary to do a detailed analysis depending on the situation.

8.7.2 Davis method

The Davis Method (Shen et al., 1981a, b) assumes a parabolic failure surface that passes through the toe of a vertical wall (Fig. 8.6). A slope stability analysis using the method of slices is used to evaluate the contribution of the nails to overall stability. Components of the tensile forces in the nails are considered parallel and perpendicular to the failure surface. Two conditions are considered in the analysis:

1. The failure surface extends beyond the reinforced zone.
2. The failure surface is entirely within the reinforced soil mass.

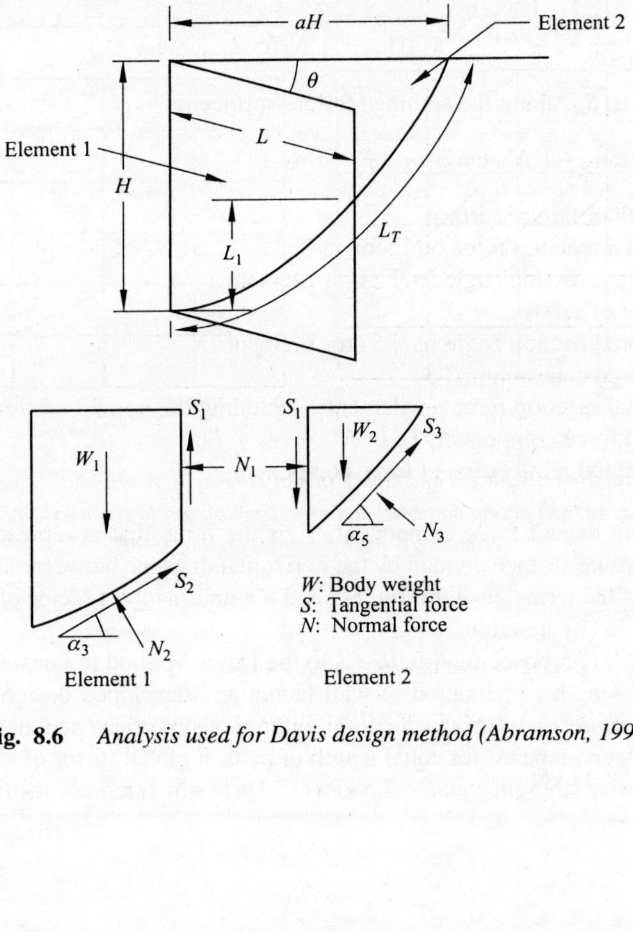

Fig. 8.6 *Analysis used for Davis design method (Abramson, 1994)*

The solutions for analysis of these two conditions contain the factors of safety and therefore must be solved by iteration. With reference to the elements 1 and 2 indicated in Fig. 8.6, N_1, N_2, N_3 and S_1, S_2 and S_3 represent the normal and shear forces acting on the elements.

The force equilibrium equations for Element 1 are:

$$N_2 = (W_1 - S_1)(\cos\alpha_3) - N_1 \sin\alpha_3 \tag{8.1}$$

$$S_2 = (W_1 - S_1)(\sin\alpha_3) + N_1 \cos\alpha_3 \tag{8.2}$$

where W_1 = weight of Element 1,
S_1 = vertical tangential force between Elements 1 and 2,
α_3 = inclination of failure surface at the base of Element 1,
N_1 = horizontal side force between Elements 1 and 2 or $\frac{1}{2}K_0(H - L_1)^2$.

The force equilibrium equations for Element 2 are :

$$N_3 = (W_2 + S_1)(\cos\alpha_5) - N_1 \sin\alpha_5 \tag{8.3}$$

$$S_3 = (W_2 + S_1)(\sin\alpha_5) - N_1 \cos\alpha_5 \tag{8.4}$$

where W_2 = weight of Element 2,
α_5 = inclination of failure surface at base of Element 2.

The total driving force S_D along the assumed failure surface is the sum of forces S_2 and S_3.

$$S_D = (W_1 - S_1)\sin\alpha_3 + (W_2 + S_1)\sin\alpha_5 + N_1(\cos\alpha_3 - \cos\alpha_5) \tag{8.5}$$

The total resisting force, S_R, along the assumed failure surface is

$$S_R = c'L_r + N_3 \tan\phi_2 + N_{2'} \tan\phi_1 + T_T \tag{8.6}$$

where, L_T = length of failure surface
N_3 = normal reaction force on Element 2
ϕ_1 = factored friction angle (ϕ/FS) for Element 1
FS = factor of safety
ϕ_2 = factored friction angle (ϕ/FS) for Element 1
c' = factored cohesion (c/FS)
N_2' = normal reaction force on element 1 including the normal reinforcement (nail) force component, T_N or $N_2', = N_2 + T_N$
T_T = tangential reinforcement force component.

The total reinforcement or nail force component, T, is the force that is representative of the length beyond the assumed failure surface divided by the horizontal distance between nails. The global factor of safety is the ratio of the terms given by Eq. 8.5 and 8.6 and involves factor of safety on either side and hence must be solved by iteration.

Elias and Juran (1991) proposed modifications to the Davis Method to consider pull-out resistance of nails, different nail lengths, inclination of wall facing and developed design charts, illustrated in Fig. 8.7. These charts represent solutions for four common geometries encountered in highway practice. The charts have been prepared for equal length of nails, a global factor of safety equal to 1.0 and a constant design cohesive strength equal to 7.5 kPa (150 psf) which represents the minimum cohesion necessary to construct stable initial cuts. In using these charts, the soil strength (ϕ) should be factored by 1.20, and the ultimate interface shear force (τ_{ult}) factored by 1.75 to 2.0 depending on the quality of the data in order to obtain structure dimensions consistent with the recommended factors of safety. To use the design charts, the pull-out resistance is defined by the non-dimensional parameter

$$S = \frac{\tau_{\text{ult}} D_g}{\gamma S_h S_v},$$

where τ_{ult} is the appropriately factored ultimate interface shear stress, D_g is the drill hole diameter, S_n and S_v are respectively the horizontal and vertical nail spacing and γ is the unit weight of soil. For an assumed design S value, the charts are used with the factored frictional strength of the *in situ* soil (ϕ) to obtain the corresponding L/H (where L and H are the length of the nails and height of the structure, respectively). The drill hole diameter and spacing can then be varied to arrive at the most economical design. The following example illustrates the use of Modified Davis charts.

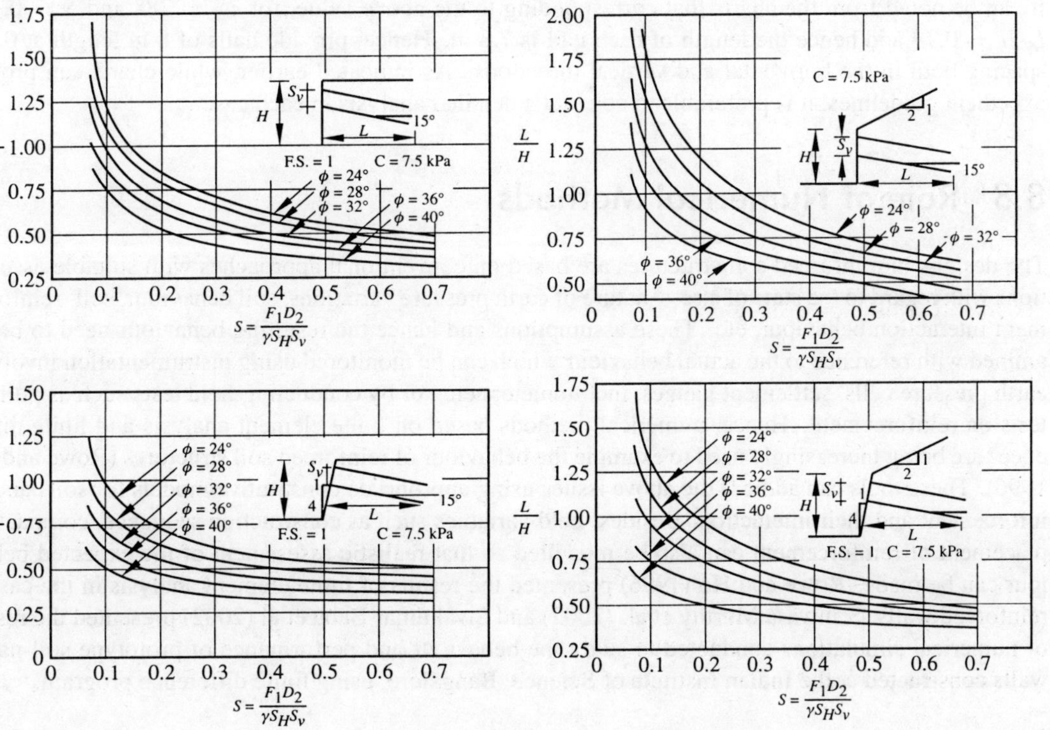

Fig. 8.7 *Modified Davis design charts (Schaefer, 1997)*

Example 8.2 In an excavation 10 m deep and inclined 14° with the vertical ($1H : 4V$), the soil nailing technique needs to be used for stabilisation. The design parameters are $\phi_d = 28°$ and bulk density $\gamma = 20\ \text{kN/m}^2$ and the vertical and horizontal spacings are both 0.5 m. A field pull-out test was conducted on a 20 mm diameter, 6 m nail driven at 9 m depth. Arrive at the length and spacing of the nails.

Solution Ribbed steel rods have been used for pullout test as well as nails. As the surface of the rods is considered to have striations, interface friction angle (ϕ_μ) is taken as equal to design friction angle (ϕ_d) of the in-situ soil.

Ultimate interface shear stress = Pull-out resistance of the nail at 9 m depth

$$= \gamma h \times L \times \pi d \tan \phi_\mu = 20 \times 9 \times 6 \times \pi \times 0.02 \times \tan 28 = 36.08 \text{ kN}.$$

Use a partial safety factor of 1.3 to get ultimate interface shear stress τ of 27.75. The non-dimensional parameter is

$$= \frac{\tau D}{\gamma S_h S_v}$$

$$S = \frac{27.75 \times 0.02}{18 \times 0.5 \times 0.5} = 0.123$$

It can be noted from the charts that corresponding to the above value, for $\phi_d = 28°$ and $S = 0.123$, $L/H = 0.78$ and hence the length of each nail is 7.8 m. Hence, provide nails of 8 m length at 0.5 m spacing both in the horizontal and vertical directions. As indicated earlier, while charts can provide expedient guidelines, it is preferable to conduct a detailed analysis in each case.

8.8 Role of Numerical Methods

The designs in reinforced soil structures are based on conventional approaches with suitable assumptions with regard to the state of stress, nature of earth pressure variations, soil behaviour, soil–reinforcement interaction behaviour, etc. These assumptions and hence the resulting behaviour need to be examined with reference to the actual behaviour which can be monitored using instrumentation involving earth pressure cells, settlement gauges, inclinometers etc., or by conducting field tests such as pull-out tests on reinforcement. Hence, numerical methods based on finite element analysis and finite differences are being increasingly used to examine the behaviour of reinforced soil structures (Rowe and Ho, 1996). These tools can address the above issues using appropriate constitutive models for soils and reinforcement and their interaction. Besides, field variables such as construction sequence, compaction, placement of reinforcement can also be modelled so that realistic assessment of the expected behaviour can be made. Rowe and Ho (1996) presented the results of finite element analysis in the case of reinforced walls. Srinivasa Murthy et al. (2002) and Sivakumar Babu et al (2002) presented the results of numerical simulations conducted to study the behaviour and performance of prototype soil-nailed walls constructed at the Indian Institute of Science, Bangalore, using finite difference program.

8.9 Behaviour in Seismic Conditions

Soil-nailed systems are inherently flexible as well as coherent and are stable in seismic conditions. Felio et al. (1990) noted that during the 1989 Loma Prieta Earthquake in the San Francisco Bay, the performance of several soil-nailed structures subjected to significant levels of shaking was very good. Field studies, numerical analysis as well as centrifuge tests showed that reinforced soil structures can withstand higher levels of shaking in earthquake conditions compared to conventional structures, which only confirms the need to adopt this technology in earthquake-prone areas.

8.10 Concluding Remarks

This chapter presents soil nailing technique in detail. The principle, the components of the technique, its behavioural aspects and design methods were discussed. There is considerable scope for increased use of the technique in Indian conditions in view of its cost effectiveness and simplicity.

Bibliography

[1] Abramson, L.W. (1994) In Situ Ground Reinforcement, *Ground Control and Improvement* (Xanthakos, L., W. Abramson and D.A. Bruce, Eds), John Wiley and Sons, Inc., New York, NY pp. 331–405.

[2] Bruce, D.A. and R.A. Jewell (1987) Soil Nailing: Application and Practice - Part I and II *Ground Engineering*, Vol. 20, No 1, pp. 10–33.

[3] CLOUTERRE (1991) French National Research Project 'Recommendations CLOUTERRE 1991—Soil Nailing Recommendations 1991, Presses de l'Ecole Nationale des Ponts et Claussees'

[4] Elias, V. and Juran, I. (1991) *Soil Nailing for Stabilization of Highway Slopes and Excavations*, United States Federal Highway Administration, Publication No. FHWA-RD-89-193, June.

[5] Felio, G.Y., Vucetic, M., Hudson, M., Barar, O., and Chapman, R. (1990) Performance of soil nailed walls during the October 17, 1989 Loma Prieta Earthquake. *Proceedings, Forty-third Canadian Geotechnical Conference*, Quebec, pp. 165–173.

[6] Gassler, G. and Gudehus, G. (1981) Soil nailing: Some mechanical aspects of in situ reinforced earth, *Proceedings of the 10th International Conference on Soil Mechanics and Foundation Engineering*, Stockholm, Sweden, 3, pp. 665–670.

[7] Gassler, G. (1993) The first two field tests in the history of soil nailing on nailed walls pushed to failure, *International Symposium in Paris*, 7–34.

[8] Gassler, G. (1996) Design of reinforced excavations and natural slopes using new European codes, Keynote lecture, *International Symposium on Earth Reinforcement* (IS Kyushu '96), Fukuoka, Japan.

[9] Rowe, R.K. and Ho, S.K.P. (1996) Some insights into reinforced wall behaviour based on finite element analysis, *Proceedings of the International Symposium on Earth Reinforcement*, Fukuoka, 485-490.

[10] Schaefer, R. (1997) ASCE Geotechnical Special Publication No. 69, *Ground Improvement, Ground Reinforcement, Ground Treatment Developments 1987–1997*.

[11] Shen, C.K., Herrmann, L.R., Romstand, K.M., Bang, S., Kim, Y.S., and Denatale, J.S. (1981a) *In-situ Earth Reinforcement Lateral Support System*, Report No. 81-03, Department of Civil Engineering, University of California, Davis.

[12] Shen, C.K., Bang, S. Romstad, J.M., Kulchin, L., and Denatale, J.S. (1981b) Field measurements of an earth support system, *Journal of the Geotechnical Engineering Division, ASCE*, 107, No. GT-12.

[13] Sivakumar Babu, G. L., Srinivasa Murthy, B.R. and Srinivas A. (2002) Analysis of construction factors influencing the behaviour of soil-nailed retaining walls, *Journal of Ground Improvement*, Vol. 6, No. 3, July 2002, 137–143.

[14] Srinivasa Murthy, B.R., G.L. Sivakumar Babu and A. Srinivas (2002) Analysis of prototype soil nailed retaining wall, *Journal of Ground Improvement*, Vol. 6, No. 3, 129–136.

[15] Stocker, M.F., Korber, G.W., Gassler, G., and Gudehus, G. (1979) Soil nailing, *International Conference on Soil Reinforcement*, Paris, 2, pp. 463–474.

9 Use of Geosynthetics for Filtration and Drainage

9.1 Introduction

Geotextile filters are rapidly replacing graded granular filters in geotechnical design. Similar to graded granular filters, their performance needs to be assured by proper design. This chapter presents the design procedure and various applications of geotextile filters in filtration, drainage and erosion control systems. The design presented is generic such that national standards and local conditions can be incorporated while evaluating the required properties of a geotextile. A step-by-step design approach is adopted and examples have been provided to emphasise the procedure.

9.2 Background

One major area of geotextile use is in drainage applications such as filters/drains in landfills, trenches, interception drains, pavement drains and permeable roadway bases. The filter restricts the movement of soil particles as water flows into the drain structure and is collected and/or transported downstream. Geocomposites consisting of a drainage core surrounded by a geotextile filter are often used as the drain in these applications. Geotextiles are also used as filters in erosion control systems.

Because of their comparable performance, improved economy, consistent properties, and ease of placement, geotextiles have successfully replaced graded granular filters in almost all drainage applications. They must thus, perform the same functions as graded granular filters in the following aspects:

- To allow water to flow through the filter into the drain, and to continue doing this throughout the life of the project; and
- To retain the soil particles in place and prevent their migration (piping) through the filter (if some soil particles do migrate, they must be able to pass through the filter without binding or clogging the downstream media during the life of the project).

Geotextiles, like graded granular filters, require proper engineering design or they may not perform as desired. Unless flow requirements, piping resistance, clogging resistance and survivability/constructability requirements are met with as specified, the geotextile filtration system may not perform properly. For this, its construction must be monitored to ensure that the materials involved are installed correctly.

9.3 Applications

Properly designed geotextiles or geotextile–soil system/geocomposites can be used as a replacement for (or in conjunction with) conventional graded granular filters in almost any drainage application. In most drainage and filtration applications, their use can be justified over conventional graded granular filter material because of the cost advantage resulting from

- use of cheaper drainage aggregate;
- possible use of smaller-sized drains;
- possible elimination of collector pipes;
- expedient construction;
- lower risk of contamination and segregation of drainage aggregate during construction; and,
- reduced excavation.

Examples of drainage applications include:

- Filters around trench drains and edge drains: to prevent soil from migrating into the drainage aggregate or the system while allowing water to exit from the soil.
- Filters beneath permeable pavement bases, blanket drains and base courses. Prefabricated geocomposite drains and geotextile-wrapped trenches are used in the construction of pavement edge drain.
- Drains for structures such as retaining walls and bridge abutments: The geotextile separates the drainage aggregate or the system the from the backfill soil, while allowing free drainage of ground water and infiltration water. Geocomposite drains are especially useful in this application.
- Geotextile wraps for slotted or jointed drain and well pipes: to prevent filter aggregate from entering the pipe, while allowing the free flow of water into the pipe.
- Interceptor, toe drains, and surface drains: to aid in the stabilization of slopes by allowing excess pore pressures within the slope to dissipate, and by preventing surface erosion. Geocomposites have been successfully used in this application.
- Chimney and toe drains for earth dams and levees: to provide seepage control.

Some examples of geotextiles use as filters in erosion control systems are listed below.

- A riprap armour stone placed over geotextile filters have been found successful in protecting precipitation and runoff collection in high-velocity diversion ditches.
- Geotextiles may be used in slope protection to prevent or reduce erosion from precipitation, surface runoff, and internal seepage or piping. In this case the geotextile may replace one or more layers of granular filter materials which would be placed on the slope in conventional applications.
- Riprap-geotextile erosion control systems may also be required along streambanks to prevent encroachment of roadways or appurtenant facilities by erosion.
- Similarly, a riprap-geotextile systems may be used for scour protection around bridge structures.

- A riprap-geotextiles system can also be effective in reducing erosion caused by wave attack or tidal variations when facilities are constructed across or adjacent to large bodies of water.
- Finally, hydraulic structures such as culverts, drip inlets, and artificial stream channels may require protection from erosion. In such applications, if vegetation cannot be established or the natural soil is highly erodible, a geotextile can be used beneath armour materials to increase erosion resistance.

Koerner (1998) provides a number of examples of these applications. Several applications may require placement of the filter layer below water. In these cases, geotextiles provide easier placement in comparison with conventional granular filter layers, and continuity of the filter medium is assured. In each of these applications flow through the geotextile is perpendicular to the plane of the fabric. In other applications such as vertical drains in soft foundation soils, lateral drains below slabs and behind retaining walls and gas transfer media, the flow may occur both perpendicular to and transverse to the plane of the geotextile. In many of these applications, geocomposite drains may be more appropriate.

All geosynthetic designs must begin with an assessment of the criticality and severity of the project conditions (see Table 9.1). First developed by Carroll (1983) for drainage and filtration applications, the concept of critical–severe projects (and thus, the level of engineering responsibility required), should be applied to all geosynthetic applications. These concepts have been reflected in the reliability based designs in which different levels of performance are quantified and targeted.

Table 9.1 *Guidelines for evaluation of critical nature of severity for filtration, drainage and erosion control applications (Carroll, 1983)*

A. Critical nature of project		
Item	Critical	Less critical
1. Risk of loss of life and/or structural damage due to drain failure	High	None
2. Repair costs versus installation costs of drain	Very high	Equal or less
3. Evidence of drain clogging before potential catastrophic failure	None	Yes
B. Severity of conditions		
Item	Severe	Less Severe
1. Soil to be drained	Gap-graded, pipable, or dispersible	Well-graded or uniform
2. Hydraulic gradient	High	Low
3. Flow conditions	Dynamic, cyclic, or pulsating	Steady state

The level of design depends on the criticality of the project and the severity of the hydraulic and soil conditions (Table 9.1). For critical projects in particular, consideration of the risks involved and the consequences of geotextile filter failure have to be assessed with great care while selecting the appropriate geotextile. Conservative designs are recommended for critical projects including the ones with severe hydraulic conditions. Geotextile selection should not be based on cost alone. The cost of the geotextile is usually minor in comparison to other components and the construction costs of a drainage system. The condition of the soil to be drained (Table 9.1) is also very important. Certain gap-graded and very well-graded soils may be internally unstable; that is, they can experience piping or internal erosion. Unstable soils include gap-graded, broadly-graded and dispersive soils such as rock flour, and other highly erodible soils. On the other hand, a soil is internally stable if it is self-filtering and if its own fine particles do not move through the pores of its coarser fraction (Lafleur, et al., 1993). Most design criteria assume that the soil to be filtered is internally stable, i.e., it will not pipe internally. If unstable soil conditions are encountered, performance tests should be conducted to select a suitable geotextile.

An evaluation of the grain size curve will provide a good indication of the stability of the soil. In gap-graded soils, coarse and fine fraction are likely to exist, with very little medium fraction. If there is an insufficient quantity of soil particles in the medium fraction, fine soil particles pipe through the coarse fraction.

A filter should ideally perform the following functions:

- It must retain the soil, which means that the size of the openings should be smaller than a specified value.
- It must remain permeable enough to allow flow, which means that the size of the openings should be larger than a specified value.

9.4 Conventional Granular Filter Design Criteria

Prior to introduction of geotextiles, granular materials were used extensively for geotechnical engineering purposes. Conventional design requires the consideration of both retention and permeability criteria which are expressed in terms of size or diameter of soil/filter particles obtained from sieve analysis. They are given by

Retention criterion: $\dfrac{d_{15}(\text{filter})}{d_{85}(\text{soil})} \leq 5$

Permeability criterion: $\dfrac{d_{15}(\text{filter})}{d_{15}(\text{soil})} > 5.$

(The upper limit of the ratio is up to 20.) In addition, $\dfrac{d_{50}(\text{filter})}{d_{50}(\text{soil})} \leq 25$ needs to be satisfied (Murthy, 1995).

9.5 Geotextile Filter Requirements

The following are the criteria that have been proposed as necessary for geotextile filters by various authors. Of them, the method proposed by Giroud (1988) and Luettich et al., (1992) are comprehensive and are discussed in the following sections. Aspects such as: (i) consideration of whether permeability or retention is the primary function of a filter within the given application; (ii) quantification of the internal stability of soil; and (iii) survivability and durability issues are presented in a systematic fashion. The various criteria are

- retention criteria,
- permeability criteria,

- anti-clogging criteria,
- survivability criteria and
- durability criteria

which are used in the selection of filter material.

A number of parameters such as boundary conditions and material properties need to be assessed in the filter design. Assessment of boundary conditions includes assessment of confining pressures and flow conditions.

9.6 Boundary Conditions

The confining pressure affects the filter behaviour, both in coarse-grained and fine-grained soils. High confining pressure affects the soil retention as well as flow properties. Flow conditions need to be evaluated in terms of flow rate and flow direction. Flow rate is essentially a function of soil hydraulic conductivity as well as hydraulic gradient. For a constant flow rate, if the changes are slow, the flow is termed steady state. If the flow rate changes are rapid, it is termed dynamic. Furthermore, the flow is both uni-directional and reverse-directional. Hence, depending on the flow rate and the flow direction, classifications are possible in terms of conditions such as steady state and dynamic state as well as uni-direction and reversible direction.

9.7 Drain and Filter Properties

A complete understanding of the properties of the drainage medium adjacent to the filter is necessary for its design. Drainage medium can be sand or a geosynthetic. For sands, particle size distribution, porosity, hydraulic conductivity and angularity are needed, while for geosynthetic drainage media, porosity, hydraulic transmissivity and thickness are the necessary criteria.

Filter medium can range from fine-grained to fine-sand to geosynthetic including woven and non-woven. For granular filter media, the particle size distribution, hydraulic conductivity and thickness, and for geosynthetic filter media, the apparent opening size, permitivity, porosity and thickness are useful data for design.

9.8 Design Criteria

9.8.1 Soil retention criteria

Soil retention is a process in which the particle movement is resisted by internal force field due to cohesion, particle size distribution, etc. In the case of granular soils, the resistance to movement is a measure of internal stability of the soil, which is a function of particle size distribution. The following are the three useful parameters for design purposes.

(a) Coefficient of uniformity, C_u, is a measure of the slope of the finer portion of the particle size distribution of a soil, defined as

$$C_u = \frac{d_{60}}{d_{10}}.$$

A soil that exhibits a wide distribution of particle sizes will have a larger C_u values, whereas a soil with uniform distribution of particle sizes will have smaller C_u values.

(b) Linear coefficient of uniformity, C'_u, is a measure of the slope of a straight line approximation of the soil particle size distribution curve. Mathematically, the following rule is applicable for a straight line on a semi-logarithmic particle size distribution.

$$C'_u = \frac{d'_{50}}{d'_0} = \frac{d'_{60}}{d'_{10}} = \frac{d'_{70}}{d'_{20}} = \cdots = \frac{d'_{100}}{d'_{50}} = \sqrt{\frac{d'_{100}}{d'_0}}$$

(c) Coefficient of curvature, C_c, is a measure of how curved the finer portion of a particle size distribution is. The coefficient of curvature is

$$C_c = \frac{(d_{30}/d_{10})}{(d_{60}/d_{30})} = \frac{(d_{30})^2}{(d_{60}/d_{10})}.$$

In the case of fine-grained soils, resistance to movement is due to cohesive nature of the soil, which is well characterized by plasticity index. Dispersion potential which indicates the tendency of the soil to disperse in the presence of water, is another indicator evaluated by double hydrometer ratio (DHR). In this test, the grain size distribution curves of the soil are determined from hydrometer test with and without a dispersing agent. The ratio corresponding to per cent finer than 0.005 mm diameter is taken as the DHR.

Determination of soil retention requirements such as particle size distribution, Atterberg limits, dispersion potential, soil density conditions indicating the effect of confining stress, were all considered and design charts prepared by Giroud (1988). These guidelines are presented in Fig. 9.1 and Fig. 9.2. Table 9.2 and Table 9.3 give details on typical relative densities and typical hydraulic gradients (Giroud, 1988).

Table 9.2 *Typical hydraulic gradients (Giroud, 1988)*

Drainage application	Typical hydraulic gradient
Standard dewatering trench	1.0
Vertical wall drain	1.5
Pavement edge drain	1
Applications in landfill	1.5
Dams	10
Inland channel protection	1
Shoreline protection	10
Liquid impoundment	10

Retention criterion is given as

$$O_{e\,(\text{geotextile})} \leq BD_{(\text{soil})} \tag{9.1}$$

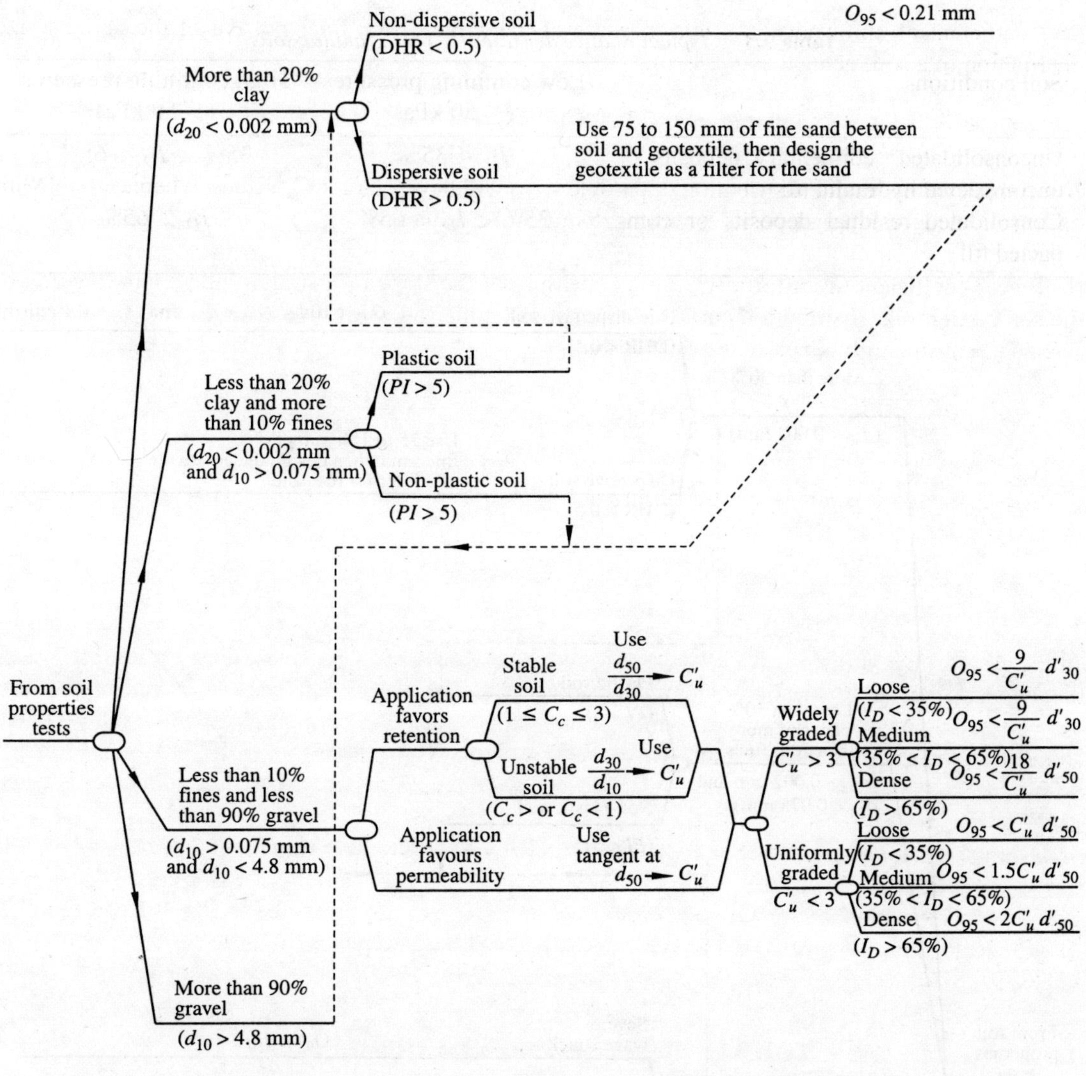

Fig. 9.1 *Chart 1 showing soil retention criteria for steady state flow conditions (Giroud, 1988)*

where O_e = effective opening size in the geotextile for which e is the percent openings that are smaller than the opening size O (mm), usually O_{90} or O_{95};

B = a coefficient (dimensionless); and

$D_{(\text{soil})}$ = representative soil particle size (mm), usually the medium to larger fractions or D_{85}.

The coefficient B generally ranges from 0.5 to 5 and is a function of the method used to evaluate the opening size (e.g. wet versus dry methods), the characteristic O_e value, type of soil to be filtered, its density, representative particle size, and uniformity coefficient C_u (if the soil granular), the type of geotextile (woven or nonwoven), and the flow conditions.

Table 9.3 *Typical relative densities (I_D) for granular soils*

Soil conditions	Low confining pressures (≤ 50 kPa)	High confining pressures (> 50 kPa)
Unconsolidated sedimentary deposits or uncompacted hydraulic fill	$I_D \leq 35\%$	$35\% < I_D < 65\%$
Consolidated residual deposits or compacted fill	$35\% < I_D < 65\%$	$I_D > 65\%$

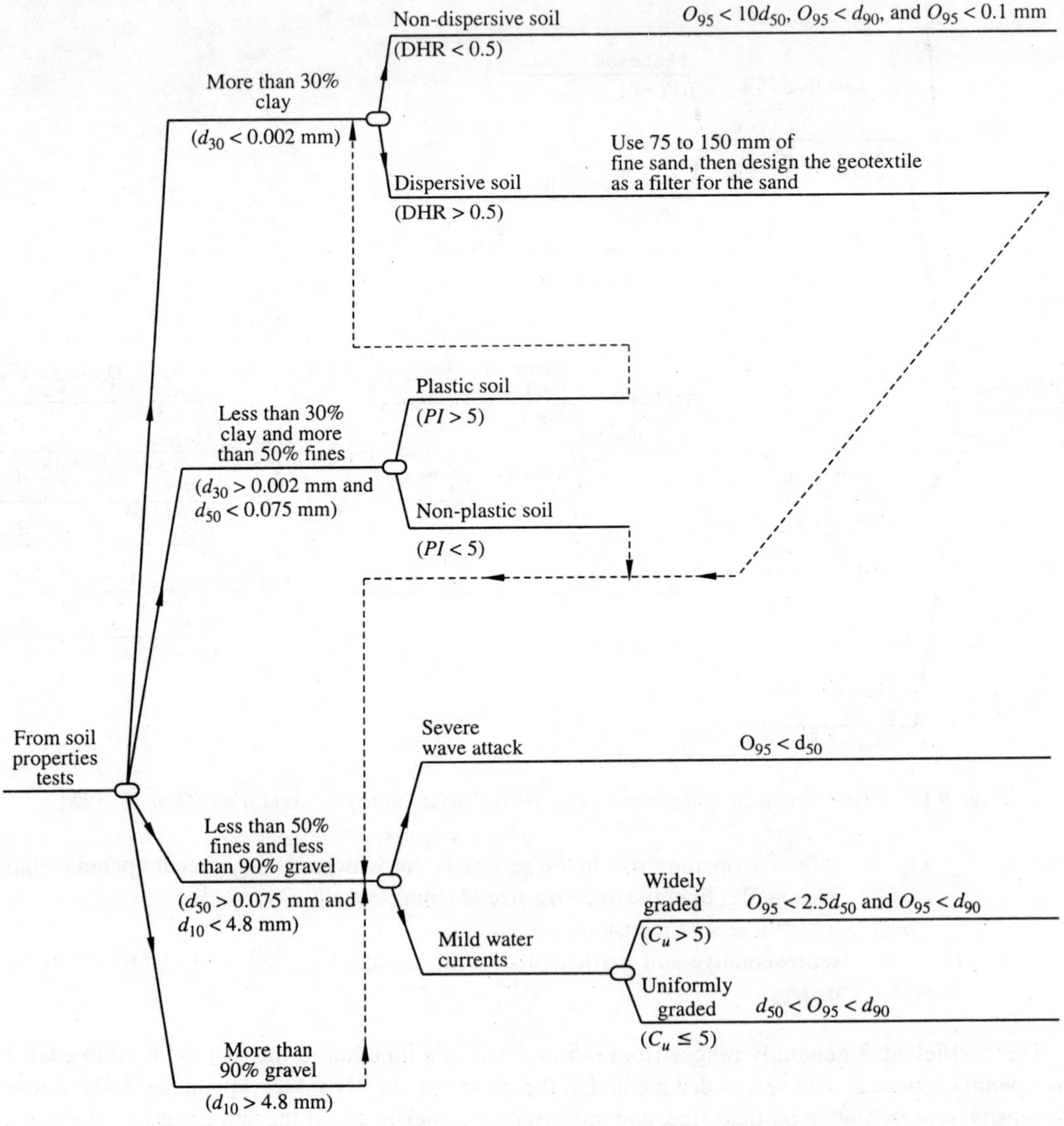

Fig. 9.2 *Chart 2 showing soil retention criteria for dynamic flow conditions (Giroud, 1988)*

9.8.2 Geotextile permeability criteria

The permeability criterion as given by Giroud is as follows:

$$k_g > i_s k_s \quad \text{where} \quad i_s > 1.0.$$

k_g is the geotextile hydraulic conductivity and i_s the hydraulic gradient in the soil next to the geotextile layer. Typical values of hydraulic gradients for different applications are given in Table 9.2. k_s is the soil hydraulic conductivity. For estimation of k_s, guidelines can be obtained from Fig. 9.3 for non-critical applications.

Permeability/permittivity criteria

The general equation for permeability is

$$k_g \geq FS \times k_s, \tag{9.2}$$

$$\psi_{\text{allow}} \geq FS \times \psi_{\text{required}}. \tag{9.3}$$

For less critical applications and less severe conditions, typically the following relation is used.

$$k_g \geq k_s \tag{9.4}$$

For critical applications and severe conditions, the factor of safety (FS) should be increased by an order of magnitude of 10 or greater, as follows.

$$k_g \geq 10k_s \tag{9.5}$$

The minimum permittivity requirement to make sure that a geotextile with a sufficient flow capacity is used is given by

$$\psi \geq 0.1 \text{ to } 1 \text{ second}^{-1} \tag{9.6}$$

In these equations,

k = Darcy coefficient of permeability (m/s); and
ψ = geotextile permittivity, which is equal to k_g/t_g (s^{-1}) and is a function of the hydraulic head. Here t_g is the geotextile thickness.

For actual flow capacity, the permeability criteria for non-critical applications is conservative, since equal quantities flow through a relatively thin geotextile in significantly less time than through a thick granular filter. Even so, some pores in the geotextile may become blocked or plugged with time. Thus, for critical or severe applications, Eq. (9.5) is recommended to provide additional conservatism.

9.8.3 Anti-clogging criteria

To minimize the risk of clogging, the following is adopted:

- Use the largest available opening size satisfying the retention criteria
- For nonwovens, porosity > 30%
- For woven geotextiles, percent open area ≥ 4%

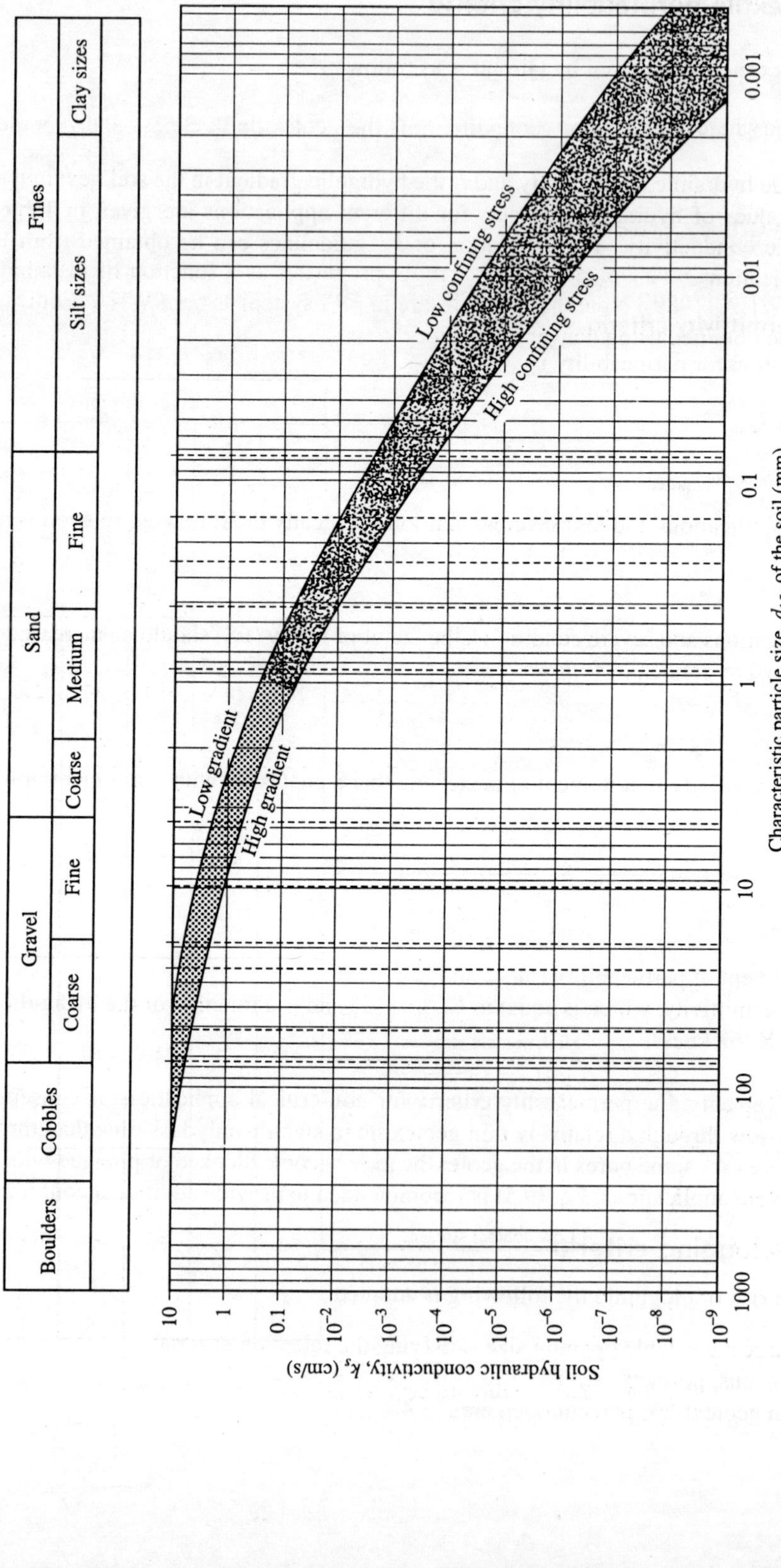

Fig. 9.3 *Relationship between soil hydraulic conductivity and characteristic particle size d_{15} (mm) of the soil (Giroud, 1998)*

The porosity of a nonwoven geotextile is given by

$$n = \left(\frac{1-\mu}{t_g \rho}\right) \times 100,$$

where μ is the mass per unit area of geotextile, t_g is the geotextile thickness and ρ is the density of filament.

9.8.4 Survivability criteria

Figure 9.4 gives the summary of geotextile strength properties as a function of installation severity obtained from FWHA (1989). Note that the units are in FPS system as per FWHA. Equivalant units in SI system may be obtained using conversion factors.

Installation condition	Contact stresses	Grab strength (lbs)	Elongation (%)	Sewn seam strength (lbs)	Puncture strength (lbs)	Burst strength (psi)	Trapezoid tear (lbs)
Moderate installation condition (Typical drainage applications)	High contact stresses (Angular drainage media) (Heavy compaction) or (Heavy confining stress)	180	N/A	160	80	290	50
Moderate installation condition (Typical drainage applications)	Low contact stresses (Rounded drainage media) (Light compaction) or (Light confining stress)	80	N/A	70	25	130	25
Severe installation conditions (Shoreline protection and armoured systems applications)	High contact stresses (Direct stone placement) (Drop height > 3 ft)	200	15	180	80	320	50
Severe installation conditions (Shoreline protection and armoured systems applications)	Low contact stresses (Sand or geotextile cushion) (Drop height < 3 ft)	90	15	80	40	140	30

Fig. 9.4 *Survivability strength requirements (FWHA, 1989)*

9.8.5 Durability criteria

Aspects such as resistance of the geotextiles to ultraviolet and adverse chemical environments need be studied for specific application. Geotextiles must be protected against extensive exposure to sunlight by incorporating anti-oxidants such carbon black or titanium oxide in their composition. Geotextiles should also be resistant to chemicals.

Example 9.1 Project specifications state that dewatering trenches be installed at every 7.6 m for the dewatering of a parking lot. The drains are at least 1.2 m below the finished grade. The drains should be constructed with open gravel wrapped in a geotextile layer. A 10 cm diameter perforated pipe is placed (as shown in Fig. E9.1). The *in situ* soil is a residual sandy silt (% fines = 51%, % clay = 4%, $PI = 0$, $C_u' = 1.7$, refer Fig. E9.2).

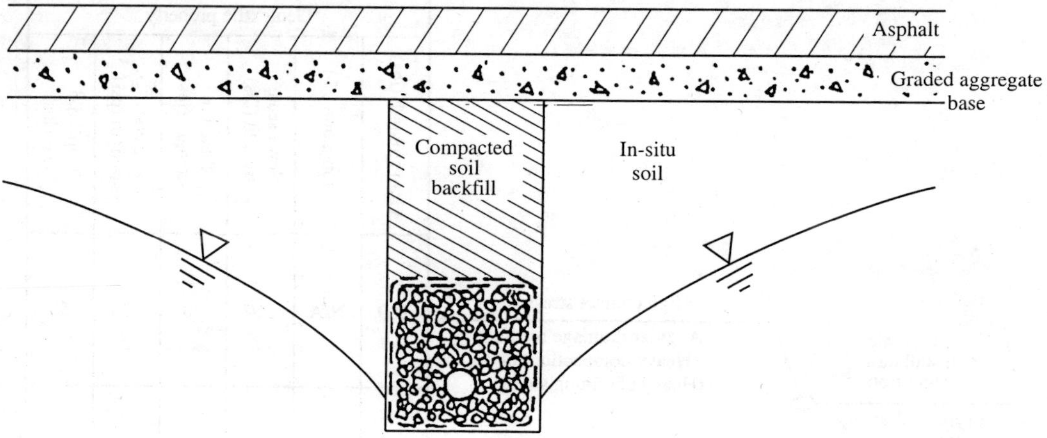

Fig. E9.1 *The soil is residual sandy silt, which classifies as an ML (sandy silt), in accordance with ASTM D2487*

Solution

(1) Soil retention criteria: The drainage medium is gravel and the application favours permeability. The confining stress effects are not significant as the drain is at shallow depth and the flow conditions are uni-directional.

From Table 9.3, I_D is in the range of 35% and 65%.

Corresponding to the above conditions, from Chart 1 (Fig. 9.1),

Geotextile apparent opening size $\leq d_{95} \leq 1.5C_u'd_{50} \leq \quad 1.5 \times 1.7 \times 0.071 \leq 0.18$ mm.

(2) Geotextile permeability criteria: From Fig. 9.3, corresponding to $d_{15} = 0.017$ mm, $k_s = 5(10^{-5})$ cm/s. From Table 9.2, typical hydraulic gradient, $i = 1.0$. Minimum allowable geotextile permeability,

$$k_g \geq i_s k_s$$

$$\geq 1.0 \times (5.0 \times 10^{-5} \text{ cm/s})$$

$$\geq 5.0 \times 10^{-5} \text{ cm/s}.$$

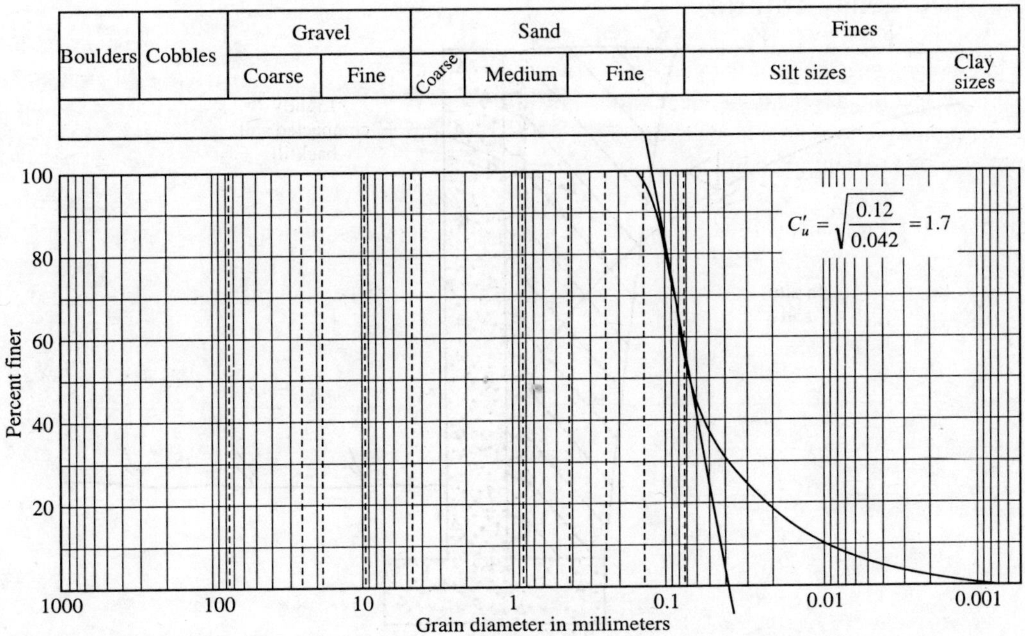

Fig. E9.2

(3) Anti-clogging requirements: From anti-clogging requirements, use the largest available O_{95}, which satisfies the retention criteria, or $O_{95} = 0.18$ mm. If the geotextile is nonwoven, porosity should be more than 30%. If the geotextile is woven, percent open area $\geq 4\%$.

(4) Survivability criteria: Choose a geotextile material that satisfies the criteria in Fig. 9.4, corresponding to moderate installation conditions and low contact stress.

(5) Durability criteria: Ultraviolet resistance is not expected to be a problem owing to the complete burial of the geotextile. In addition, chemical resistance requirements are not serious as the permeant is uncontaminated ground water.

Example 9.2 For a continuous retaining wall of 3 m height, the project specifications require that a continuous wall drain be installed on the backside of the wall (Fig. E9.3). The drain is to be constructed with either gravel or geosynthetic drainage material with a geotextile filter. The retaining wall has lightly compacted backfill, classified as silty sand and its properties are read from Fig. E9.4.

Solution

The confining stress can be considered to be low due to the shallow depth of installation, and the flow conditions are uni-directional. Further, the following soil properties are known. (Percent fine = 32%, per cent clay = 7%, $PI = 0$).

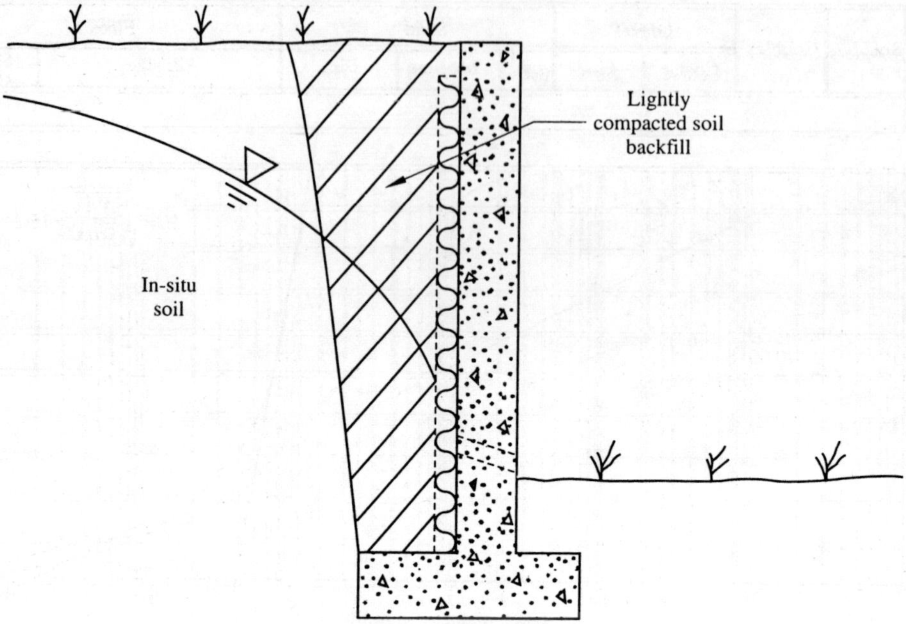

Fig. E9.3 *The soil (lightly compacted soil) backfill classifies as an SM (silty sand)*

(1) Soil retention criteria: Assuming that the application favours permeability,

$$C'_u = \sqrt{\frac{d_{60}}{d_{10}}} = \sqrt{\frac{3.4}{0.013}} = 16.2$$

From Chart I (Fig. 9.1), $I_D < 35\%$ and

$$O_{95} \leq (9/C'_u)d'_{50} \leq (9/16.2)0.22 \leq 0.11 \text{ mm.}$$

(2) Geotextile permeability requirements: From Fig. 9.3, for $d_{15} = 0.02$ mm, $k_s = 5(10^{-5})$, and from Table 9.2, $i = 1.5$. Minimum allowable geotextile permeability

$$k_g \geq i_s k_s$$
$$\geq 1.5 \times (5.0 \times 10^{-5} \text{ cm/s})$$
$$\geq 7.5 \times 10^{-5} \text{ cm/s.}$$

(3) Anti-clogging requirements: Use the largest available O_{95} that satisfies the retention criteria. If the selected geotextile is nonwoven, the porosity should be more than or equal to 30%. If it is woven, the open area should be more than 4%.

(4) Survivability requirements: Use Fig. 9.4 and consider that the installation conditions are moderate and have a low contact stress.

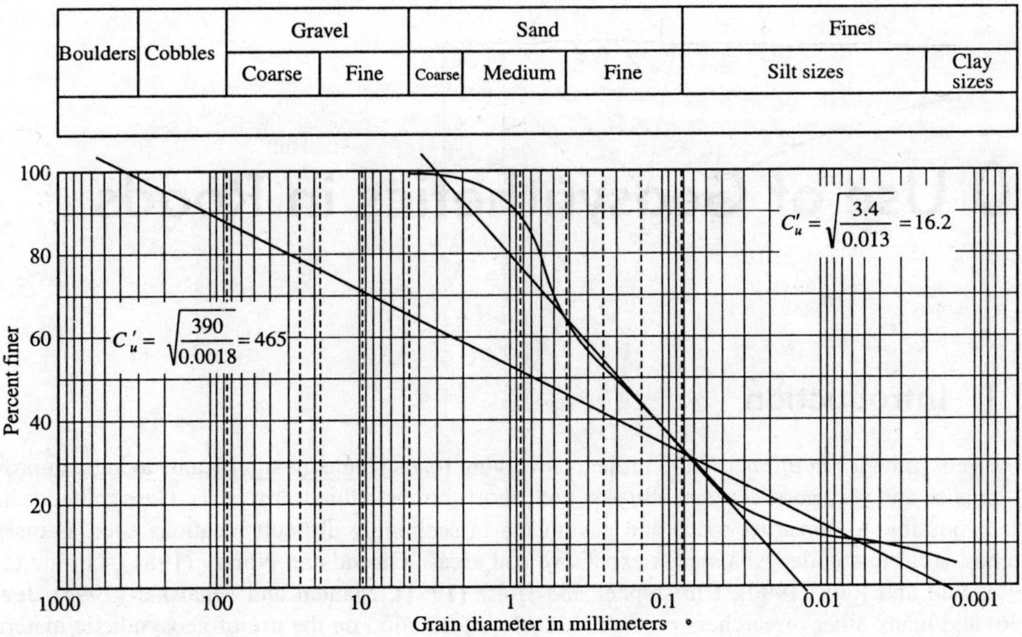

Fig. E9.4

(5) Durability requirements: Similar to the earlier problem.

Bibliography

[1] Carroll, R.G., Jr. (1983) Geotextile filter criteria: Engineering fabrics in transportation construction, *Transportation Research Record 916*, Transportation Research Board, Washington, D.C., January, pp. 46–53.

[2] FHWA (1989) *Design of Riprap Revetment*, Hydraulic Engineering Circular No. 11, Federal Highway Administration.

[3] Giroud, J.P. (1988) Review of geotextile filter design criteria, *Proceedings of First Indian Conference on Reinforced Soil and Geotextiles*, Calcutta, 1988, pp1-6.

[4] Koerner, R.M. (1998) *Designing with Geosynthetics*, 4th Edition, Prentice Hall, New Jersey.

[5] Lafleur, J., Mlynarek, J. and Rollin, A.L. (1993) Filter criteria for well graded cohesionless soils, Filters in geotechnical and hydraulic engineering, *Proceedings of the First International Conference—Geo-Filters*, Karlsruhe, Brauns, Schuler, and Heibaum Eds., Balkema, pp. 97–106.

[6] Luettich, S.M., J.P. Giroud and R.C. Bachus (1992) Geotextile filter design guide, *Geotextile and Geomembranes*, Vol. 11, 355–70.

[7] Murthy, V.N.S. (1995) *A Text Book of Soil Mechanics and Foundation Engineering*, Sai Kripa Technical Consultants Publicaton.

10 Use of Geosynthetics in Roads

10.1 Introduction

Use of geosynthetics in infrastructure projects involving roads results in significant savings, improved performance and very good serviceability on both short-term and long-term basis. Geosynthetics have made it possible to construct roads and pavements in seemingly difficult locations such as marshy stretches, soft/organic deposits and in expansive soil areas. Giroud and Noiray (1981), Giroud et al. (1994), Rao and Raju (1990), Christopher and Holtz (1991), Mandal and Mhaiskar (1994), Jewell (1996) and many other researchers provide valuable information on the use of geosynthetic materials in construction of pavements.

10.2 Geosynthetics in Roadways

A geosynthetic placed at the interface between the aggregate and the subgrade functions acts as a separator to prevent two dissimilar materials (subgrade soils and aggregates) from intermixing (Fig. 10.1). Geotextiles and geogrids perform this function by preventing penetration of the aggregate into the subgrade (localised bearing failures). In addition, geotextiles prevent intrusion of subgrade soil up into the base course aggregate. Localised bearing failures and subgrade intrusions only occur in very soft, wet and weak subgrades. A small amount of fines can significantly reduce the friction angle of granular aggregate to low values. Therefore, separation is important to maintain the design thickness, the stability and the load-carrying capacity of the base course. Soft subgrade soils are most susceptible to disturbance during construction activities such as clearing, grubbing, and initial aggregate placement. Geosynthetics can help minimize subgrade disturbance and prevent loss of aggregate during construction. Thus the primary function of the geotextile in this application is separation, and can in some cases be considered a secondary function for geogrids.

The overall performance of the road, even if is designed based on the separation criteria, is also influenced by secondary functions of filtration, drainage, and reinforcement. The geotextile acts as a filter to prevent fines from migrating into the aggregate due to high pore water pressures induced by dynamic wheel loads. It also acts as a drain, allowing the excess pore pressures to dissipate through the geotextile. Hence subgrade soils gain strength through consolidation and improve with time.

Geosynthetics provide reinforcement through three possible mechanisms.

- Lateral restrainment of the base and subgrade through friction and the interlocking of the aggregate, soil and the geosynthetic (Fig. 10.2*a*).

- Increase in the system bearing capacity by forcing the potential bearing capacity failure surface to develop along alternate, higher shear strength surfaces (Fig. 10.2*b*)
- Membrane support of the wheel loads (Fig.10.2*c*)

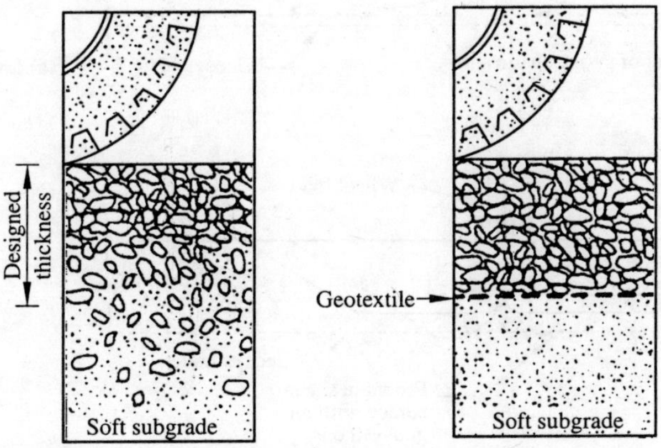

Fig. 10.1 *Concept of geotextile separation in roadways*

When an aggregate layer is loaded by a wheel or track, the aggregate tends to move laterally (Fig. 10.2*a*) unless it is restrained by the subgrade or geosynthetic reinforcement. Soft, weak subgrade soils provide very little lateral restraint, so when the aggregate moves laterally, ruts develop on the aggregate surface and also in the subgrade. A geogrid with good interlocking capabilities or a geotextile with good frictional capabilities can provide resistance to lateral aggregate movement. Another possible geosynthetic reinforcement mechanism is illustrated in Fig. 10.2*b*. Using the analogy of a wheel load to a footing, the geosynthetic reinforcement forces the potential bearing capacity failure surface to follow an alternate higher-strength path. This tends to increase the bearing capacity of the roadway. A third possible geosynthetic reinforcement function is a membrane-type support of the wheel loads, as shown conceptually in Fig. 10.2*c*. In this case, the wheel load stresses must be high enough to cause plastic deformation and ruts in the subgrade. If the geosynthetic has a sufficiently high tensile modulus, tensile stresses will develop in the reinforcement, and the vertical component of this membrane stress will help support the applied wheel loads. As tensile stresses within the geosynthetic cannot be developed without some elongation, a wheel path rutting (in excess of 100 mm) is required to develop the membrane-type support. Therefore, this mechanism is generally limited to temporary roads or to the first aggregate lift in permanent roadways.

10.3 Applications

10.3.1 Temporary and permanent roads

Roads and highways are broadly classified into two categories: permanent and temporary, depending on their service life, traffic applications, or desired performance (John, 1987). Permanent roads include both paved and unpaved ones, which usually remain in service 10 years or more. Permanent unpaved

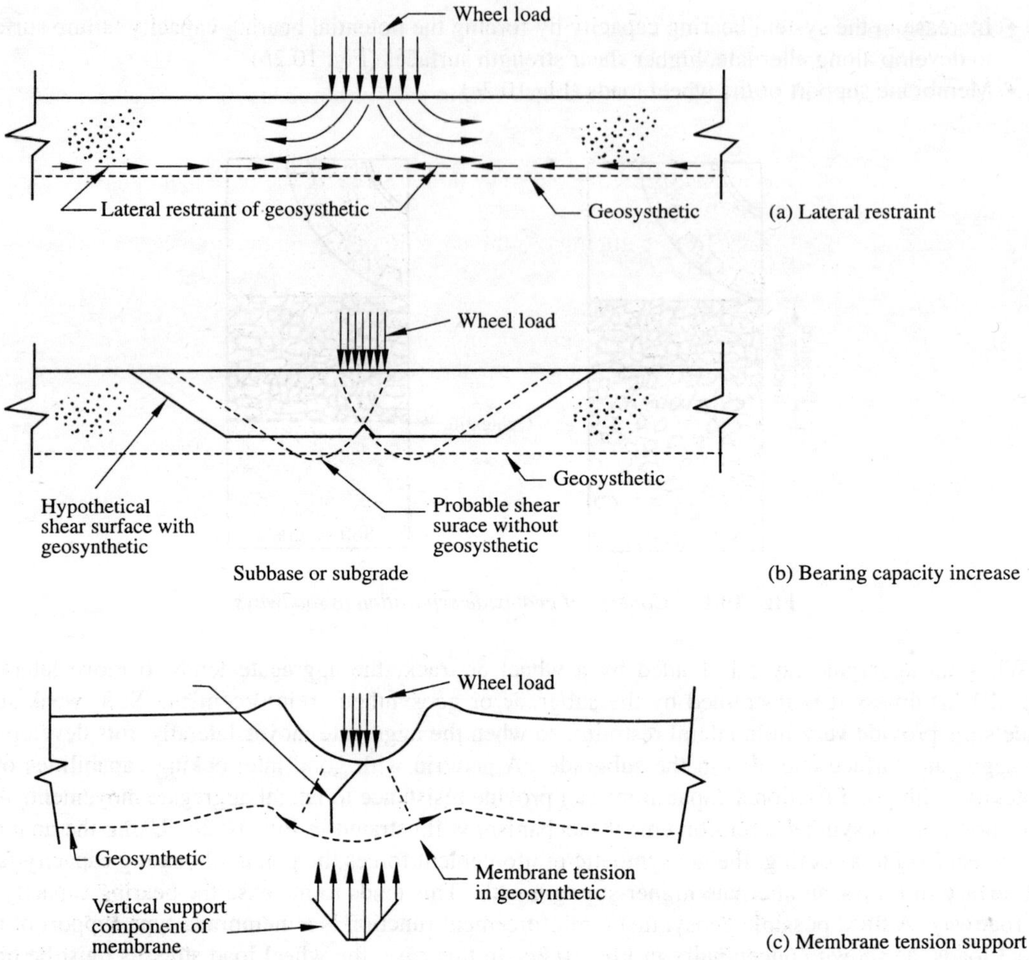

Fig. 10.2 *Reinforcement functions of geosynthetics in roadways (Haliburton et al., 1981)*

roads may be subjected to more than a million load repetition. Temporary roads remain in service for only short periods of time (often less than 1 year), and are usually subjected to fewer than 10,000 load applications during their service lives. Temporary roads include detours, haul and access roads, construction platforms, and stabilised working areas required for the construction of permanent roads, as well as embankments over soft foundations.

Geosynthetics allow access for construction equipment to sites where the soil is normally too weak to support the initial construction work. This is one of the more important uses of geosynthetics. In some cases, even if it is possible that the finished roadway can be supported by the subgrade, it may be virtually impossible to begin construction of the embankment or roadway. Such sites require dewatering, excavation and replacement with select granular materials, utilization of stabilisation aggregate, chemical stabilisation, etc. Geosynthetics can often be a cost effective alternate to these expensive foundation treatment procedures.

10.3.2 Benefits

Geosynthetics used in roadways on soft subgrades perform several functions for enhancing stability, as listed below.

1. Reducing the intensity of stress on the subgrade and preventing the base aggregate from penetrating into the subgrade (function: separation)
2. Preventing subgrade fines from pumping or otherwise migrating up into the base (function: separation and filtration).
3. Preventing contamination of the base materials, which may allow more open-graded, free-draining aggregates to be considered in the design (function: filtration).
4. Reducing the depth of excavation required for the removal of unsuitable subgrade materials (function: separation and reinforcement)
5. Reducing the thickness of aggregate required to stabilize the subgrade (function: separation and reinforcement)
6. Reducing the disturbance of the subgrade during construction (function: separation and reinforcement)
7. Allowing an increase in subgrade strength over time (function: filtration)
8. Reducing the differential settlement of the roadway, which helps in maintaining pavement integrity and uniformity (function: reinforcement). Geosynthetics also aid in reducing differential settlement in transition areas from the cut to the fill.
9. Reducing maintenance and extending the life of the pavement (functions: all)

Geosynthetics are also used in permanent roadways to provide capillary breaks to reduce frost action in frost-susceptible soils, and to provide membrane-encapsulated soil layers (MESL) to reduce the effects of seasonal water content changes on swelling clays.

10.4 Role of Subgrade Conditions

Geotextile separators have been in use for more than 20 years in the stabilization of very soft wet subgrades. Based on practical experience and several case histories summarized by Haliburton, Lawmaster, and McGuffey (1981) and Christopher and Holtz (1991), the following subgrade conditions are considered to be the most appropriate for geosynthetic use in roadway construction. The classification of soil is based on United Soil Classificaiton Systems (USCS) and American Association of State History and Transport Officials (AASHTO).

- Poor soils
  (USCS: SC, CL, CH, ML, MH, OL, OH, and PT)
  (AASHTO: A-5, A-6, A-7-5, and A-7-6)
- Low undrained shear strength
  $\tau_f = C_u < 90$ kPa
  California Bearing Ratio (CBR) < 3
  Resilient modulus (M_R) ≈ 30 MPa
- High water table
- High sensitivity

Under these conditions, geosynthetics function primarily as separators and filters to stabilise the subgrade, improving construction conditions and allowing long term strength improvements in the subgrade. If large ruts develop during the placement of the first aggregate lift, then some reinforcing effect is also present. As a general recommendation, geotextile functions given in Table 10.1 are appropriate for the corresponding subgrade strengths.

Table 10.1 *Effectiveness of geosynthetics as a function of subgrade strength*

Undrained shear strength (kPa)	Subgrade CBR	Functions
60–90	2–3	Filtration and possibly separation
30–60	1–2	Filtration, separation, and possibly reinforcement
< 30	< 1	All functions, including reinforcement

The table suggests that geotextiles do not provide any useful function when the undrained shear strength is greater than about 90 kPa (CBR about 3%). If traffic is limited, as it is in many temporary roads, or if shallow (75 mm) ruts are acceptable as in most construction operations, then a maximum undrained shear strength of about 90 kPa (CBR = 3%) for geosynthetic use in highway construction seems reasonable.

10.5 Design

There are two main approaches to the design of temporary and unpaved roads. The first assumes no reinforcing effect of the geotextiles, that is, the geotextile acts as a separator only. The second approach considers a possible reinforcing effect due to the geotextile. It appears that the separation function is more important for thin roadway sections with relatively small live loads where ruts, approximately 50 to 100 mm, are anticipated. In these cases a design which assumes no reinforcing effect is generally conservative. On the other hand, for large live loads on thin roadways where deep ruts (> 100 mm) may occur, and for thicker roadways on softer subgrades, the reinforcing function becomes increasingly more important if stability is to be maintained.

Based on both theoretical analysis and empirical (laboratory and full-scale field) tests on geotextiles, Steward, Williamson and Mohney (1977) determined that a certain amount of rutting would occur under various traffic conditions, both with and without a geotextile separator and for a given stress level acting on the subgrade. They present this stress level in terms of bearing capacity factors, similar to those commonly used for the design of shallow foundations on cohesive soils. These factors and conditions are given in Table 10.2.

Design approaches for unpaved roads incorporating geotextiles have been proposed by Giroud and Noiray (1981), Milligan et al, (1989a, b), Jewell (1996) and others. Among them the approach proposed by Giroud and Noiray (1981) is popular and is presented in the following sections.

Table 10.2 *Bearing capacity factors for different ruts and traffic conditions both with and without geotextile separators (Steward, Williamson, and Mohney, 1977)*

Design	Ruts (mm)	Traffic (Passes of 80 kN axle equivalents)	Bearing capacity factor
Without geotexitle	< 50	> 1000	2.8
	> 100	< 100	3.3
With geotextile	< 50	> 1000	5.0
	> 100	< 100	6.0

10.6 The Giroud and Noiray Approach

The equivalent contact area of a wheel on the road surface is taken as width $B \times$ length L. The relationships between these dimensions, the axle load P, the tyre pressure p_t and the type of vehicle are given in the following equations.

(a) Normal highway vehicles including lorries:

$$B = \sqrt{P/p_t}; \quad L = 0.707B$$

(b) Heavy construction plant vehicles with wide or double tyres:

$$B = \sqrt{1.414P/p_t}; \quad L = 0.5B$$

From Fig. 10.3*a*, the stress p applied to the subgrade by the axle is

$$p = \frac{P}{2(B + 2h\tan\alpha)(L + 2h\tan\alpha)}. \tag{10.1}$$

The analysis is not very sensitive to the exact value of $\tan\alpha$ and experiments indicate that $\tan\alpha$ lies between 0.5 and 0.7. So $\tan\alpha$ may be taken as 0.6.

Hence,

$$p = \frac{P}{2(B + 1.2h)(L + 1.2h)}. \tag{10.2}$$

Giroud and Noiray make use of the net elastic bearing capacity (q_c) and the ultimate or plastic bearing capacity (q_p), defined as:

$$q_e = \pi c_u$$
$$q_p = (\pi + 2)c_u$$

where c_u is the undrained cohesion of the underlying soil.

The significance of these two values is demonstrated by the graph of load versus settlement for the centre of a loaded plate (Fig. 10.3*b*). To control any contamination (intermixing of the aggregate with subgrade material) of the aggregate, it is suggested that in the absence of a geotextile, the applied load from the axle be limited to q_e. Substituting $p = q_e$ in Eq. (10.2) gives:

$$\pi c_u = \frac{P}{2(B + 1.2h)(L + 1.2h)}. \tag{10.3}$$

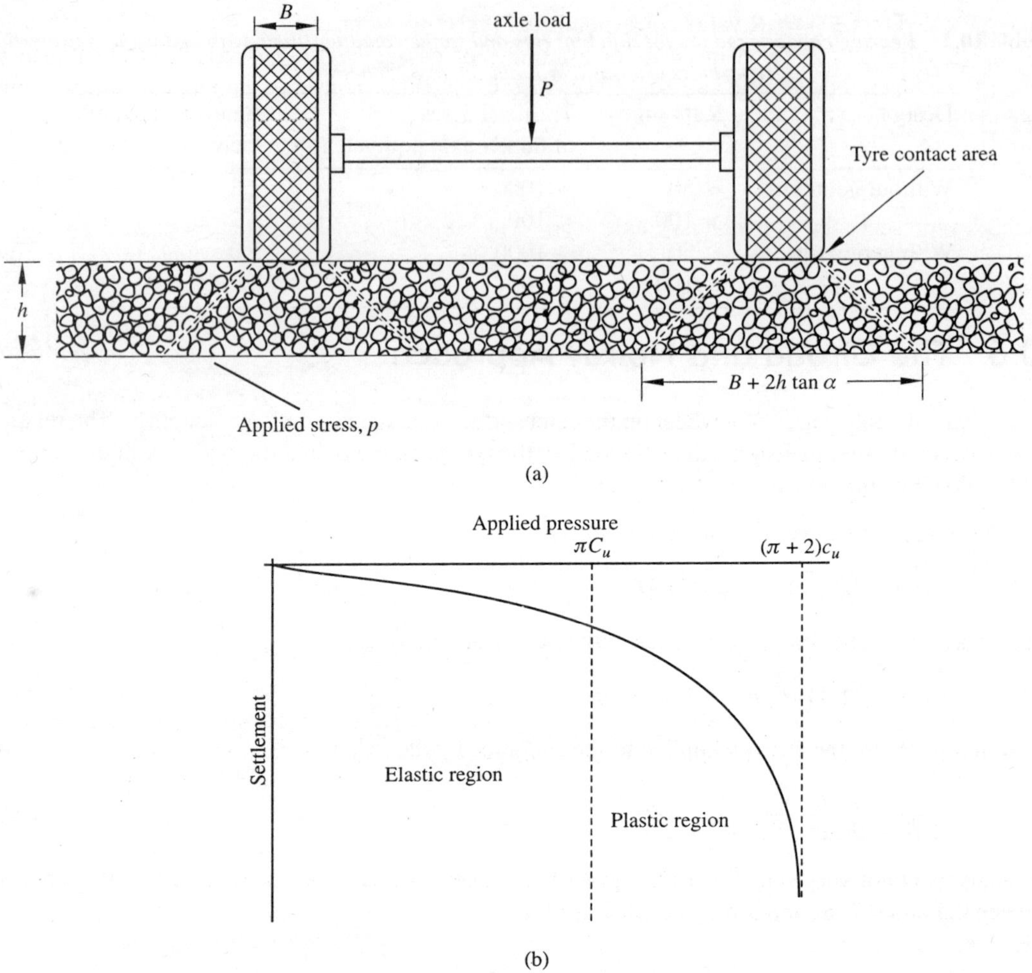

Fig. 10.3 *(a) Notation for Giroud and Noriay analysis; (b) settlement at the centre of a loaded plate*

The value of h obtained from this equation is denoted as h_0 and is the minimum depth of aggregate required for one pass of the axle, without the presence of a geotextile separator. The value of h_0 remains valid for very light traffic, up to about 20 axle passes. However, the aggregate depth must be increased to h_0 for heavier traffic using

$$h_0 = \frac{[(125 \log N - 294(r - 0.075)]}{c_u^{0.63}}, \tag{10.4}$$

where N is the number of passes of a standard axle (80 kN),
r is the rut depth in metres,
c_u is undrained soil cohesion in $\mathrm{N/m^2}$ (not in $\mathrm{kN/m^2}$), and
h_0 is the aggregate depth in meters.

It may be noted that 1% CBR value corresponds to 30 kN/m^2 (John, 1987).

If the loading is expressed in terms of a number of passes (N') of an axle load other than the standard axle load, then Giroud and Noiray suggest that it can be converted into an equivalent number of standard axle load passes (N) using

$$N/N' = (P'/P)^{3.93} \tag{10.5}$$

where P' represents the non-standard axle load. The suitability of this equation is doubtful and an alternative theoretically appropriate equation for the conversion of axle loads is given by (John, 1987, Jewell, 1996)

$$N/N' = (P'/P)^{6.2}. \tag{10.6}$$

Giroud et al. (1984) suggests that the following three factors contribute to the extra stability arising from the presence of a geotextile in an unpaved road section.

(a) Enhanced confinement of the subgrade soil
(b) Greater spread of the loading
(c) An uplift force due to the geotextile tension.

In the absence of a geotextile, the local shear (which occurs when the elastic limit for the bearing capacity is exceeded) would lead to large deformations, which in turn would accelerate the deterioration of the roadway. Confinement of the subgrade soil by a geotextile controls the local shear, enabling the design to be based on the plastic (or ultimate) bearing capacity. The openings in woven and nonwoven geotextile sheets are sufficiently small to guarantee this confining effect. In the case of geogrids, some of the benefit may be lost if a sufficient quantity of subgrade soil passes through the apertures into the road base aggregate.

If a geogrid interlocks with the road base aggregate, then the effective aggregate stiffness is increased and the axle load is distributed over a greater area. A limited number of tests by Giroud et al. (1984) on geotextiles suggest that $\tan\alpha$ increases from the standard value of 0.6 to between 0.8 and 1.0 if contamination of the aggregate is ignored. By assuming that the geosynthetic deforms to a parabolic shape, Giroud and Noiray derived the following expression for the geosynthetic uplift force (F_g) given by

$$F_g = \frac{J\varepsilon}{a\sqrt{(1 + (a/2S)^2)}}, \tag{10.7}$$

where, J is the tensile stiffness of the geosynthetic
ε is the geosynthetic strain,
a is $1/2(B + 1.2h)$ and
S is the settlement beneath the tyre.

The force F_g reduces the load P and hence the corresponding relationship is given by

$$(\pi + 2)c_u = \frac{P}{2(B + 2h\tan\alpha)(L + 2h\tan\alpha)} - \frac{J\varepsilon}{a\sqrt{1 + (a/2S)^2}}. \tag{10.8}$$

Subsequent work by Giroud et al. (1984) indicates that the benefit from the geotextile uplift force is negligible when the rut depth is 75 mm or less and a reduction of 10% or less in road base is obtained when the rut depth is 150 mm. This suggests that the confining effect becomes a primary mechanism for enhancing performance using geosynthetics.

Due to the confining effect, the ultimate bearing capacity is given by

$$(\pi + 2)c_u = \frac{P}{2(B + 1.2h_G)(L + 1.2h_G)}, \tag{10.9}$$

where h_G is the required fill depth with geotextile. The savings in aggregate depth due to the presence of geotextile (Δh) is given by

$$\Delta h = h_o - h_G.$$

The following example illustrates the calculation procedure.

Example 10.1 An unpaved road 5 m wide is to be subjected to 1000 passes of a heavy vehicle (two axles) (each axle load 120 kN). On each axle of truck, there are four 0.2 m wide tyres, inflated to a pressure 700 kN/m^2. The underlying soil has an undrained cohesion of 30 kN/m^2 and the road aggregate has unit weight of 18 kN/m^3. If the acceptable rut depth is 150 mm, determine the road thickness using Giroud and Noiray approach.

Solution

$$N' = 1000 \times 2 = 2000$$

We have

$$\frac{N}{N'} = (\frac{P'}{P})^{6.2}, \quad N = 2000(\frac{120}{80})^{6.2} = 24705.6.$$

Hence the loading is equivalent to 24705.6 passes of standard load of 80 kN.

We have

$$h_0 = \frac{125 \log N - 294(r - 0.075)}{c_u^{0.63}}$$

$$= \frac{125 \log 24705.6 - 294(0.150 - 0.075)}{(30,000)^{0.63}} = 0.796 = 0.80 \text{ m}.$$

For construction plant vehicle with wide or double tyres,

$$B = \sqrt{1.414P/p_t} \text{ and } L = 0.5B. \text{ Hence,}$$

$$B = \sqrt{1.414 \times \frac{120}{700}} = 0.50 \text{ m and } L = 0.25 \text{ m}.$$

Elastic bearing capacity

$$\pi c_u = \frac{P}{2(B + 1.2h_e)(L + 1.2h_e)}$$

$$94.24 = \frac{120}{2(B + 1.2h_e)(L + 1.2h_e)}$$

$$= \frac{120}{2(0.50 + 1.2h_e)(0.25 + 1.2h_e)}$$

where h_e is the thickness of the road corresponding to maximum pressure in elastic region in Fig. 10.3.
Substitution of the quantities leads to a quadratic equation in h_e. Solving for h_e gives

$$h_O = 0.357 \text{ m} = 0.36 \text{ m}.$$

Ultimate bearing capacity

As indicated earlier, the bearing capacity can be increased to $(\pi+2)c_u$ with the introduction of geosynthetic material (Fig. 10.3).

$$(\pi + 2)c_u = \frac{P}{2(B + 1.2h_G)(L + 1.2h_G)}.$$

Substitution of the appropriate quantities leads to h_G value of 0.22 m.

$$\Delta h = 0.36 - 0.22 = 0.14.$$

Hence depth with geosynthetic = 0.8 − 0.14 = 0.66 m.
It may be noted that savings in the aggregate to the extent of 17.5% (0.14/0.80) are obtained using geosynthetic material.

Contribution due to tension of the geosynthetic

Consideration of geosynthetic uplift force due to tension increases the ultimate bearing capacity by the term

$$\frac{J\varepsilon}{a\sqrt{1 + (a/2S)^2}}$$

as explained earlier. In the above example, if the stiffness is 100 kN/m and strain is 10% corresponding to a settlement of 10 cm, the thickness with geosynthetic reduces to 0.20 m. This can be worked out as follows. Equation (10.8) gives

$$5.14 \times 30 = \frac{120}{2(0.50 + 1.2h_G)(0.25 + 1.2h_G)} - \frac{100 \times 0.10}{1/2(0.50 + 1.2h_G)\sqrt{1 + \left[\dfrac{1/2(0.50 + 1.2h_G)}{2 \times 0.10}\right]}}$$

One value of h_G such that both right and left terms of the above equation become approximately equal by trial and error can be obtained and calculations show that $h_G = 0.20$ m is reasonably satisfactory.

$$\text{Hence } \Delta h = 0.36 - 0.20 = 0.16.$$

Hence, the depth with geosynthetic = 0.80 − 0.16 = 0.64 cm. It can be observed that the geosynthetic reinforcement contributes to a reduction of 0.02 m in the present example. Hence it is apparent that the contribution from tension of the reinforcement is less. The other criteria such as constructability and survivability of the geotextile are more important; the relevant specifications are given in the following sections.

10.7 Geotextile Survivability

Selection of a geotextile for either permanent or temporary roads depends upon the survivability criteria. These survivability requirements are not based on any systematic research but on the properties of geotextiles which have performed satisfactorily as separators in temporary roads and in similar applications. However, in the absence of any other information, they should be used as minimum property values. Table 10.3 gives an example of how survivability requirements are followed. Geotextile survivability for major projects should be verified by conducting field tests under site-specific conditions.

Table 10.3 *Physical property requirements (AASHTO 1990)*

Survivability level	Grab strength ASTM D 4632 (N) < 50% Geotextile elongation	Grab strength ASTM D 4632 (N) > 50% Geotextile elongation	Puncture resistance ASTM D 4833 (N) < 50% Geotextile elongation	Puncture resistance ASTM D 4833 (N) > 50% Geotextile elongation	Tear strength ASTM D 4632 (N) < 50% Geotextile elongation	Tear strength ASTM D 4632 (N) > 50% Geotextile elongation
High (Class 1)	1400	900	500	350	500	350
Moderate (Class 2)	1100	700	400	250	400	250
Low (Class 3)	800	500	300	180	300	180

Additional requirements	*Test method*
Apparent opening size	
1 < 50% soil passing 0.075 mm sieve, AOS < 0.6 mm	
2 > 50% soil passing 0.075 mm sieve, AOS < 0.3 mm	ASTM D 4751
Permeability	
k of the geotextile > k of the soil	
(permittivity × the nominal geotextile tthickness)	ASTM D 4491
Ultraviolet degradation	
At 500 hours of exposure, 50% strength retained	ASTM D 4355
Geotextile acceptance	ASTM D 4759

Notes:
[1]For the index properties, the first value of each set is for geotextiles which fail at less than 50% elongation, while the second value is for geotextiles which fail at greater than 50% elongation. Elongation is determined by ASTM D 4632.
[2]Values shown are minimum roll average values. Strength values are in the weakest principal direction.
[3]The values of the geotextile elongation do not relate to the allowable consolidation properties of the subgrade soil. These must be determined by a separate investigation.

10.8 Application in Paved Roads

Appropriate locations for placements of geotextiles in a paved road structure are:

(a) at the interface between the granular sub-base and the subgrade soil,
(b) near the underside of the wearing course,
(c) beneath the surface overlay to a damaged pavement.

Subgrade to sub-base interface

Placement of geotextile at this level reduces the possibilities of rutting and to some extent reduces the thickness of the sub-base, besides controlling the contamination of the sub-base by fines from the subgrade soil.

Pavement wearing course

The advantage of using a geotextile within a wearing course helps in:

(a) restriction of the reflection cracking,
(b) reduced rutting,
(c) restriction of fatigue cracking.

In areas subject to high contact stresses, such as heavily trafficked areas, taxiways, the effect is beneficial. Extensive research carried out in UK clearly indicates that use of geotextiles and geogrids considerably reduces rutting.

Pavement overlays

Geosynthetics can be used as alternatives to stress-relieving granular layers, seal coats, rubberized asphalts, etc. for controlling surface moisture infiltration and retarding reflection cracks in pavement overlays. Properly installed, asphalt-saturated geotextiles function as a moisture barrier that protect the underlying pavement structure from further degradation due to ingress of surface water. In addition, geosynthetics can provide cushioning for the overlay, thus functioning as a stress-relieving interlayer. When properly installed, both functions combine to extend the life of the overlay and the pavement section.

Bibliography

[1] American Association of State Transportation and Highway Officials, (1990) Standard Specifications for Geotextiles—M288, Standard Specifications for Transportation Materials and Methods of Sampling and Testing, AASHTO, Washington, DC., pp. 689–692.

[2] Christopher, B.R. and Holtz, R.D., (1991) Geotextiles for subgrade stabilisation in permanent roads and highways, *Proc. of Geosynthetics '91*, Atlanta, Georgia, V 2, pp. 701–713.

[3] Giroud, J.P. and Noiray, L., (1981) Geotextile reinforced unpaved roads, *Journal of the Geotechnical Engineering Division*, American Society of Civil Engineers, V 107, NGT 9, pp.1233–1254.

[4] Giroud, J.P., Ah-line, C. and Bonaparte, R. (1985) Design of unpaved roads and trafficked areas with geogrids, *Proc. Symp. Polymer Grid Reinforcement in Civil Engg,* 9–12 London.

[5] Haliburton, T.A., Lawmaster, J.D. and McGuffey, V.C., (1981), Use of engineering fabrics in transportation related applications, FHWA, DTFH61-80-C-00094.

[6] Jewell, R.A. (1996) *Soil Reinforcement with Geotextiles*, CIRIA Publication, UK.

[7] John, N.W.M. (1987) *Geotextiles*, Blackie Publishers, Glasgow and London.

[8] Milligan, G.W.E, Jewell, R.A, Houlsby, G.T and Burd, H.J (1989a) A new approach to the design of unpaved roads-Part I, *Ground Engineering*, April, 25–29.

[9] Milligan, G.W.E, Jewell, R.A, Houlsby, G.T and Burd, H.J (1989b) A new approach to the design of unpaved roads-Part II, *Ground Engineering*, November, 37–42.

[10] Steward, J.E., Williamson, R. and Mohney, J. (1977) *Earth Reinforcement*, Chapter 5 on guidelines for use of fabrics in construction and maintenance of low volume roads, US Forest Service, Portland.

11 Geosynthetics in Landfills

11.1 Introduction

While engineers, scientists and environmentalists have long cautioned decision makers about the dangers of ground water pollution caused by contaminated waste products, it is only recently that the governmental, political and public sectors have taken them seriously and begun to act. The urgency of the situation is real and today the issue attracts as much attention as any hotly debated topics like the defence, the budget and political events. Several industrialised countries have promulgated rules and regulations in this area in the form of "minimum technology guidance" that include use of geosynthetic materials to some extent. In India, the Ministry of Environment and Forests issued guidelines (MoEF, 2000) that have to be followed by all municipalities and corporations of the country to ensure that impact of municipal solid waste (MSW) is lessened to some extent. These guidelines suggest the use of geosynthetic materials in the construction of landfills. This chapter presents the advantages of using these materials in landfill construction and remediation of contaminated sites.

The containment of solid and liquid waste materials is one of the greatest societal problems that exists today. It is very clear that improper disposal and/or improper containment of waste products is a major threat to the groundwater and surface water supplies. Waste materials may contain substances that are harmful to the environment. It is essential that the waste be handled in such a way that any contamination of the ground as well as that of ground water is prevented. The primary task to achieve this is to design, construct and operate landfills so as to provide effective barriers against contamination. A schematic view of the barrier concept (Gartung, 1996) is presented in Fig. 11.1. The waste body is surrounded by a liner at the bottom and cover at the top and sides, and rests on a hard stratum. All these components act as barriers and help contain the waste. Gradients are provided within the waste body along the boundaries in such a way that that the liquid contained in the waste (called leachate, generated in the landfill due to reactions within the waste body as well as unwanted infiltration/ingress from the top and sides) is collected in a leachate system. The leachate is subsequently treated using waste water treatment methods.

Although considerable research work has been undertaken in this area during the last few years, no general and all-encompassing rational computational design method has been evolved as yet. However landfill design and construction cannot be delayed for the above reason. To ensure a certain level of safety with respect to environment, responsible national agencies have issued some basic rules for the location of landfills, their design and construction, and other related requirements, based on these research findings and available experience. The guidelines issued by the Environmental Protection Agency (EPA) of USA and similar bodies in countries like Germany, France, and the Ministry of

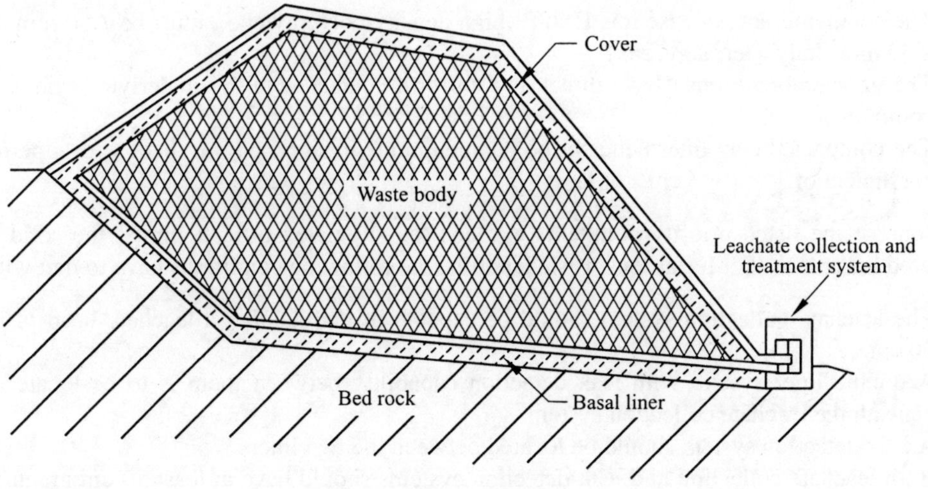

Fig. 11.1 *Barrier concept in landfills (Gartung, 1996)*

Environment and Forests of India, etc., all advocate the use of geosynthetics. In the following sections typical designs of landfills, available guidelines and the role of geosynthetics are discussed.

11.2 Typical Designs of Landfills

Considerable information on the design of landfills exists in research documents, guidelines and case studies (Daniel, 1987, Landreth, 1997, Daniel and Koerner, 1993a, b, 1997, Scuero, 1997 and others). These studies clearly demonstrate the advantage of using geosynthetics in landfill construction. Some information from the above is provided in the following sections.

11.2.1 Landfill liners

Solid waste materials that are subsequently landfilled are subdivided into non-hazardous and hazardous categories based on their chemical characteristics and site-specific leachate. Leachate is the liquid contained in the waste and the subsequent rain water that flows through it. The leachate has the representative characteristics of the solid waste and is therefore always waste-specific. The leachate is analyzed by conventional analytical techniques (toxicity characteristics leaching procedure (TCLP) sampling, mass spectroscopy, and gas chromatography analysis) and if all of the listed priority pollutants are less than the stipulated official limits of agencies such as EPA, the waste mass is considered as non-hazardous. In such a case, the most frequent design must meet the following conditions:

- A leachate collection system should be located above the liner system.
- The leachate collection system should be capable of maintaining a leachate head of less than 30 cm.
- The liner system should be a single composite liner, a double liner system with leak detection capability is not required as with hazardous waste materials.
- The single composite liner must be a geomembrane placed over a compacted clay liner.

- The geomembrane, in case it is HDPE (high density polyethylene), must be 1.50 mm (USA) to 2.50 mm (Italy, Germany, etc.)
- The geomembrane must have direct and uniform contact with the underlying compacted soil component.
- The compacted clay liner beneath the geomembrane must be 60 cm thick with a permeability coefficient of 1×10^{-7} cm/s or less.

If any one of the listed priority pollutants exceeds the stipulated official limits, the solid waste is considered hazardous. For hazardous waste landfills, the following conditions have to met with.

- The leachate collection system should be capable of maintaining a leachate head of less than 30 cm.
- A double liner system with leak detection capability between them is to be located directly beneath the leachate collection system.
- A leak detection system should be located between the two liners.
- Both leachate collection and leak detection systems should have at least 30 cm granular layers which are chemically resistant to the waste and leachate, with a permeability coefficient not less than 1×10^{-2} cm/s, or an equivalent synthetic drainage material, e.g., a geonet or geocomposite.
- The minimum bottom slope of the facility should be 0.5%.
- The leachate collection system should have a granular filter or synthetic geotextile above the drainage layer to prevent clogging.
- Both collection systems, when made of natural soils, should have a network of interconnected perforated pipes to remove the leachate, and the pipes should have sufficient strength and chemical resistance to perform under anticipated landfill loads.
- The geomembranes made of HDPE must be 1.50 mm to 2.50 mm according to the regulations existing in a country.
- The compacted clay liner beneath the secondary geomembrane must be 60 cm thick and have permeability coefficient of 1×10^{-7} cm/s or less.

11.2.2 Covers

Covers of waste landfill must perform several functions in addition to keeping the water out of the contaminated material. They minimize and control precipitation runoff, keep away the waste from plants and animals, discourage intrusion, intentional or accidental, and control gas release. The cover, or closure of a landfill is critically important since its performance must be assured over an extremely long lifetime. Lifetimes well beyond the 30 years post closure care period are often considered. Within the cover system are the following elements which must be evaluated and designed according to site-specific and waste-specific considerations.

- Vegetative cover and top soil
- Cover soil
- Surface water drainage system
- Composite barrier system (geomembrane with clay liner or geomembrane with geosynthetic clay liner)
- Gas venting layer (required for municipal solid waste)
- Final compacted cover soil over the solid waste mass

11.3 Guidelines

11.3.1 EPA guidelines

In USA, two separate sets of guidelines exist for municipal solid wastes (MSW) under 40 CFR 258 (Code of Federal Regulations) called the subtitle D regulations, and for hazardous waste under 40 CFR 264 and 265 called the subtitle C regulations (Sharma and Lewis, 1994).

Municipal solid waste (MSW) landfills

To ensure protection of health and environment, minimum criteria are specified nationwide, which applies to all existing and new MSW landfills. Besides the location criteria, the criteria for design of liners and their closure are specified in subtitle D program (40 CFR 258). The design criteria suggest that the landfills should be constructed with a composite liner and the leachate collection system is constructed to maintain less than 30 cm depth of leachate over the liner. The composite liner must consist of an upper component of flexible membrane liner (FML), with a minimum of 30 mil (1 mil = 1/1000 of an inch) HDPE material and a lower component of minimum 60 cm thick compacted soil with permeability coefficient of not more than 1×10^{-7} cm/s. The landfill units must have a final cover system designed to minimise infiltration and erosion. The erosion layer must be of 15 cm thick earthen material, capable of sustaining the growth of native plants. The infiltration layer must be of at least 45 cm thick earthen material that has permeability coefficient of less than or equal to 1×10^{-5} cm/s. Figure 11.2 shows the liner and cover system for an MSW landfill. It may be noted that the flexible membrane line (FML) or geomembrane in the cover prevents rainfall infiltration into the waste.

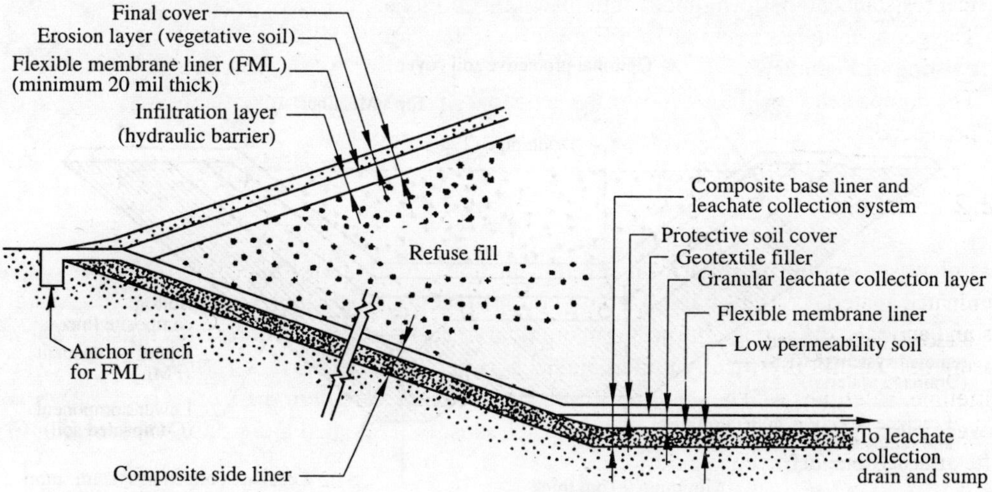

Fig. 11.2 *Typical liner and final cover systems of a MSW landfill as per EPA guidelines (Sharma and Lewis, 1994)*

For the containment of hazardous liquid waste, often called surface impoundment, there must be two or more liners and a leachate collection and removal system between such liners. The liner system must have two flexible membrane liners (FML) as shown in Fig. 11.3. The requirements for hazardous solid waste are similar and a sump is provided (Fig. 11.4) for leachate collection at the monitoring point, called the compliance point. To close the hazardous waste landfill, a multi-layered cover system shown in Fig. 11.5 is recommended.

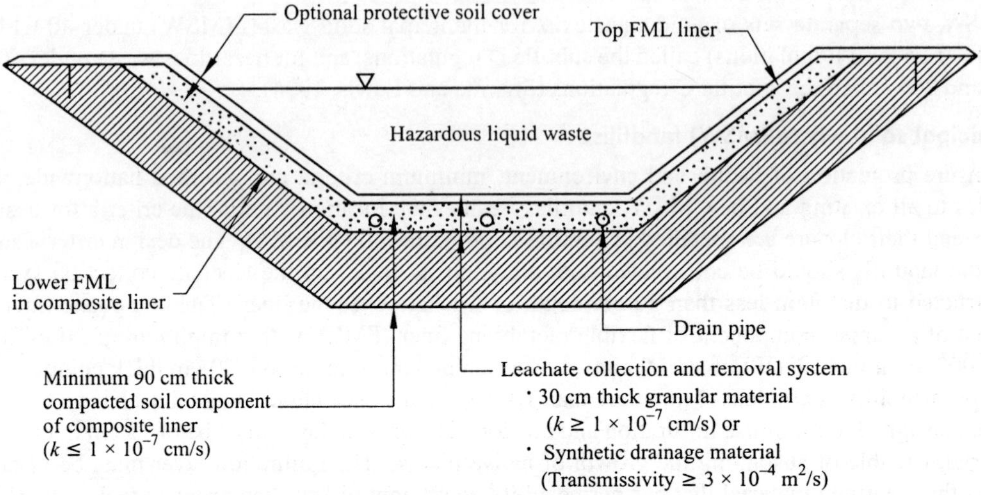

Fig. 11.3 *Schematic diagram of composite liner for surface impoundment for disposal of hazardous liquid waste (Sharma and Lewis, 1994)*

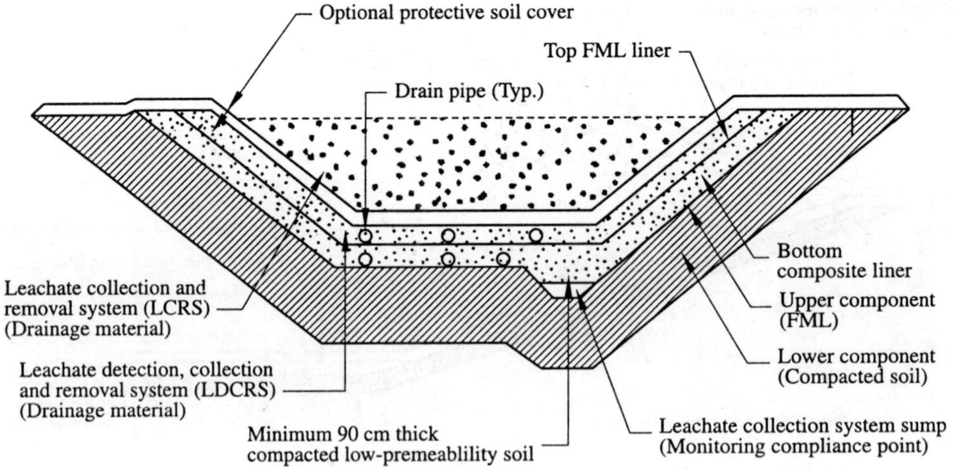

Fig. 11.4 *Schematic diagram of a double liner and leachate collection system for hazardous solid waste landfill (Sharma and Lewis, 1994)*

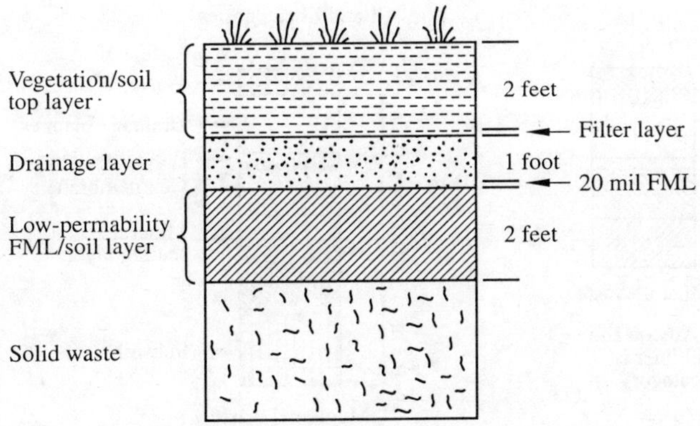

Fig. 11.5 *A schematic diagram of cover systems for closure of hazardous waste (Sharma and Lewis, 1994)*

11.3.2 Guidelines followed in Germany

In Germany three categories of landfills are identified, one for mineral or inert waste which is similar to harmless construction debris and other similar items, and the other two for municipal and hazardous wastes (Gartung, 1996). Refuse with the lowest potential are assigned to landfill category I. Materials such as municipal waste, wastes from incineration which do not have organic carbon content of more than 5% by weight go to category II landfill. Wastes that contain harmful substances exceeding the above criteria and are considered hazardous in terms of other criteria go to the category III landfill. It is important to realise that the hazardous material should be treated to the maximum extent to remove all the hazardous material before it is placed in the landfill. The lining and cover systems for these systems are shown in Fig. 11.6 and Fig. 11.7. A schematic of the leachate collection pipes at the bottom of the landfills is shown in Fig. 11.8. It may be noted that similar criteria are followed in other European countries like Switzerland and Austria.

11.3.3 Ministry of Environment and Forests (MoEF) guidelines

In India, Ministry of Environment and Forests (MoEF) and the Central Pollution Control Board (CPCB) made regulations in 2000 which deal with the criteria for location; site selection; site investigation; planning and design; waste acceptance; landfill liner and cover; construction and operation as well as inspection; and monitoring and record keeping. In the landfill liner and cover criteria, requirements of the liner and cover are specified.

MSW landfills

Guidelines suggest that composite lining systems should be provided at the base of the waste disposal area to prevent pollution of underground water resources. A 1.5 mm thick HDPE geomembrane or an equivalent overlaying a 90 cm of soil (fine-grained soil or an amended soil) having a coefficient of permeability not greater than 1×10^{-7} cm/s serves as the composite liner. Figure 11.9 shows the sections for cover and liner systems recommended for MSW landfills.

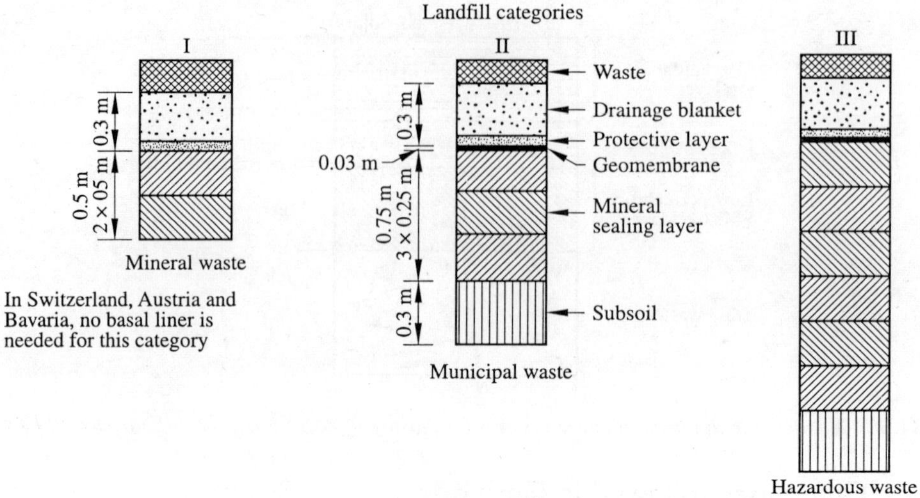

Fig. 11.6 *Typical sections for liners of landfills of different categories in Germany (Gartung, 1996)*

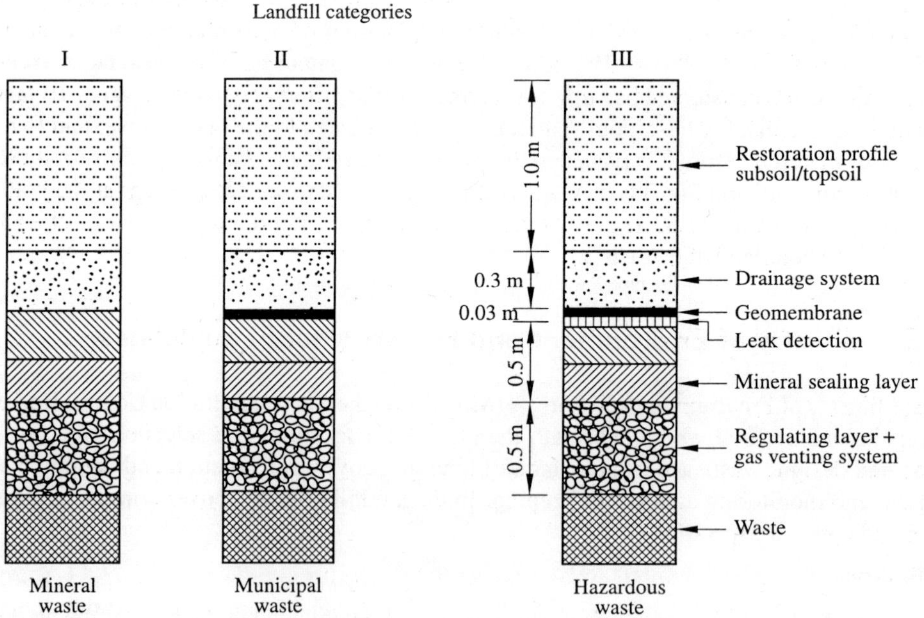

Fig. 11.7 *Typical sections for covers of landfills of different categories in Germany (Gartung, 1996)*

Hazardous waste landfills

The components of hazardous waste landfills are shown in Fig. 11.10. A leachate collection layer of thickness 30 cm or more, with a coefficient of permeability $\geq 10^{-2}$ cm/s rests on a composite liner

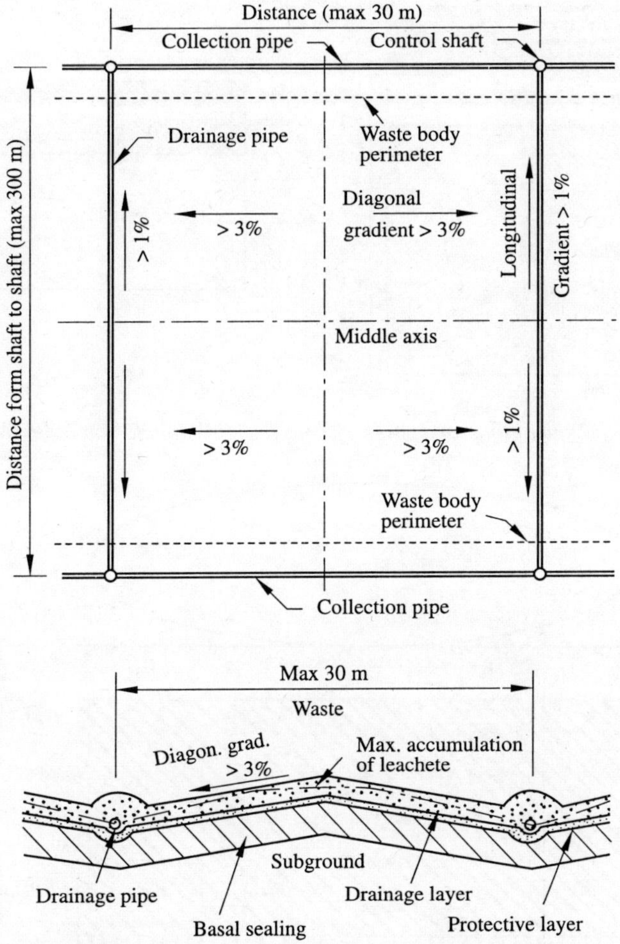

Fig. 11.8 *A schematic diagram showing network of leachate collection pipes at the bottom of landfill (Gartung, 1996)*

consisting of a HDPE geomembrane of thickness 1.5 mm overlying 150 mm thick compacted clay having a coefficient of permeability $\leq 10^{-7}$ cm/s. In regions where the rainfall is high, and the subsoil highly permeable and the water table within 2 m to 6 m beneath the base of the landfill, the liner system is a double composite liner and has two geomembranes. This is shown in Fig. 11.11. Construction requirements suggest that the geomembranes be laid in intimate contact with the compacted clay and be properly joined/welded at the seams, and that they should not be punctured by construction vehicles/tools. To ensure proper performance of geomembrane, quality control tests such as (i) thickness tests, (ii) density tests, (iii) strength tests, (iv) toughness tests, (v) durability tests (vi) chemical resistance tests (vii) field seam strength tests (viii) overlap check tests, etc., should be performed. Covers of landfills consisting of a geomembrane of thickness 1.5 mm or more and a compacted clay are required to function as a composite barrier (Fig. 11.12).

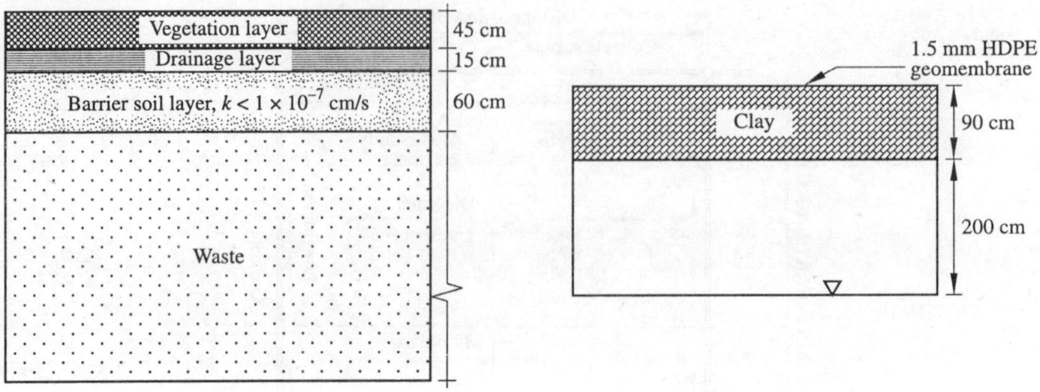

Fig. 11.9 *Typical liner and cover systems for MSW landfills in India (MoEF, 2000)*

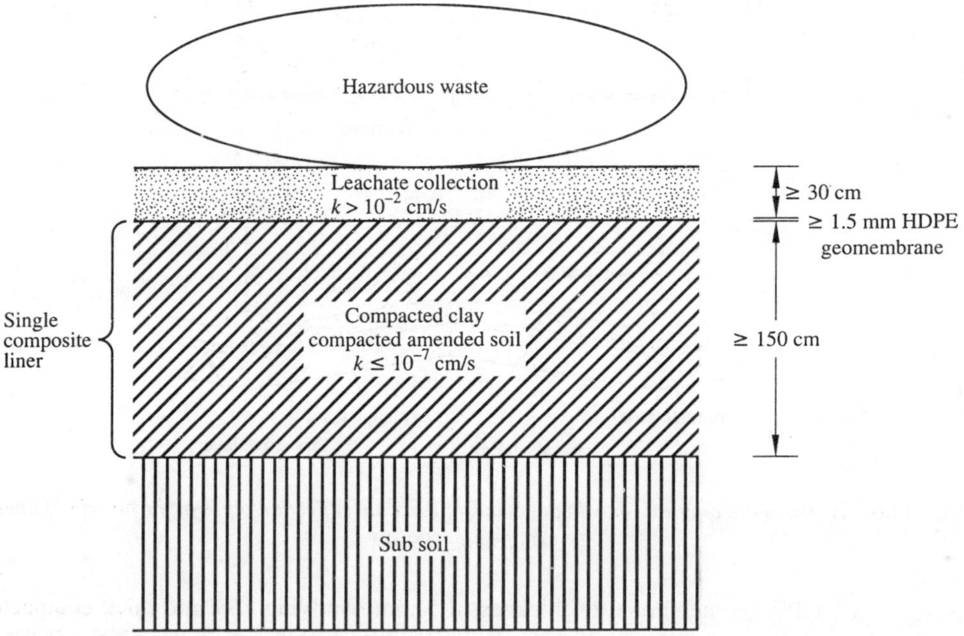

Fig. 11.10 *A schematic diagram of single composite liner system for hazardous waste landfills in India (MoEF, 2000)*

11.4 Role of Geosynthetics

Geosynthetics of every type play a key role in providing safe and secure containment of waste materials. Geomembranes, as the primary barrier, are augmented by various drainage materials such as geonets, geocomposites and geopipe. These in turn utilize geotextiles as protectors, separators, gas

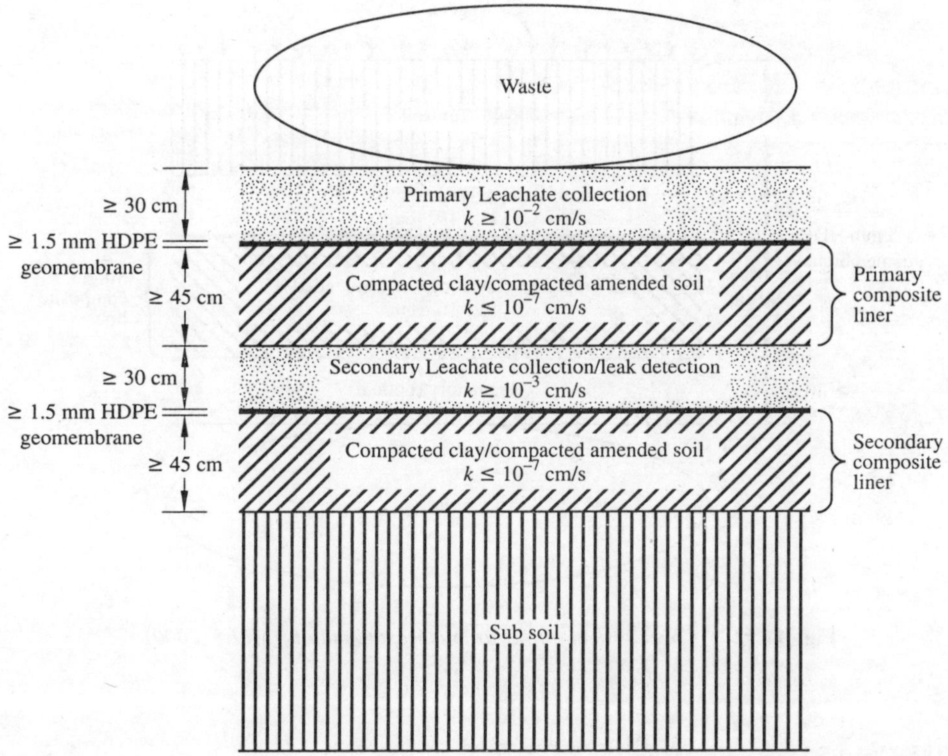

Fig. 11.11 *A schematic diagram of double composite liner system for hazardous waste landfills in the areas of port drainage and high rainfall (MoEF, 2000)*

drains and (along with geogrids) reinforcement material. Geosynthetic clay liners are used to replace blankets of natural clays. Figure 11.13 shows a typical landfill with geosynthetics for different functions. There is a large and ever growing body of technical information on geosynthetics (for example, Koerner, 1998). Among the functions of geosynthetics that are useful in a solid waste landfill, the most important types are described below. Also, see Table 11.1.

- Geomembranes (GM) are very low-permeable polymeric membrane liners or barriers used to contain the generated leachate and control fluid migration.
- Geotextiles (GT) are planar, permeable polymeric materials comprised solely of textiles used for filtration, drainage, separation, or reinforcement.
- Geosynthetic clay liners (GCL) are factory-manufactured hydraulic barriers consisting of a layer of bentonite clay or other very-low-permeability material supported by geotextiles and/or geomembranes, mechanically held together by adhesives, needling, or stitching.
- Geonets (GN) are net-like sets of interconnected polymer ribs used for the transmission of liquids in the plane of the structure.
- Geogrids (GG) are grid-like sets of interconnected polymer ribs used for reinforcement of soils or solid waste.
- Geopipes (GP) are polymeric pipes used to convey the leachate.

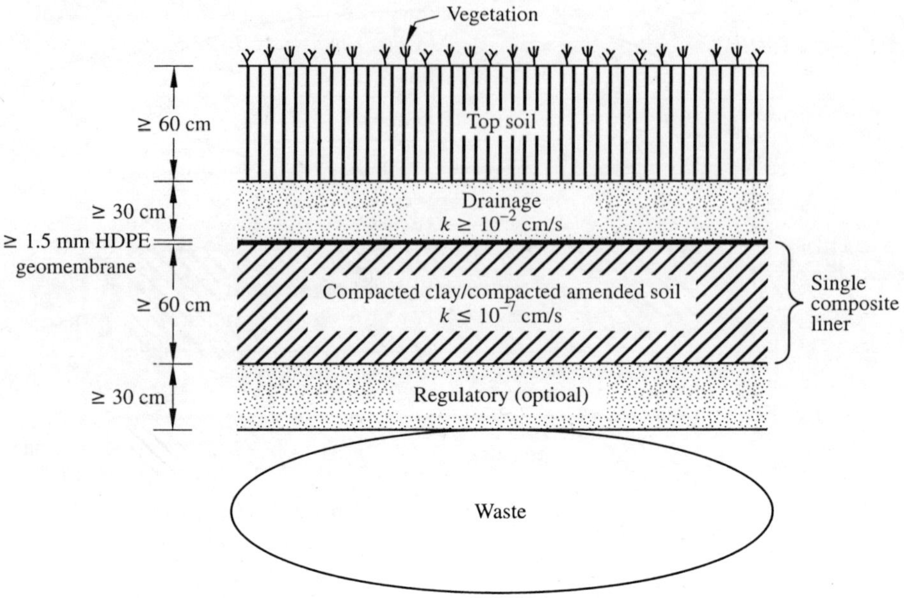

Fig. 11.12 *Cover system for hazardous waste landfills (MoEF, 2000)*

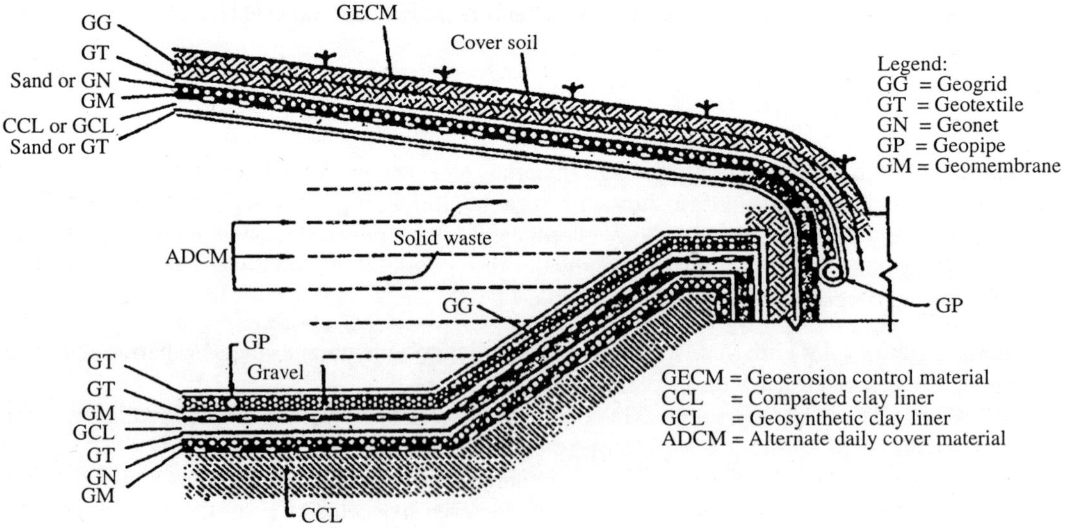

Fig. 11.13 *Various geosynthetic components that can be used in a double liner landfill system (Koerner, 1998)*

Table 11.1 *Use and primary functions of various geosynthetic materials (Refer to notation and shown in Fig. 11.13)*

Layer	Location	Geosynthetic	Primary function
1	Above waste	GT	Filtration (surface water)
2	Above waste	GC	Drainage (surface water)
3	Above waste	GM	Barrier (surface water entering, landfill gases leaving)
4	Above waste	GT	Drainage (landfill gases)
5	Below waste	GT	Filtration (leachate)
6	Below waste	GN	Drainage on side slopes (leachate)
7	Below waste	GT	Filtration for perforated pipes (leachate)
8	Below waste	GT	Separation (protection)
9	Below waste	GM	Barrier (leachate)
10	Below waste	GT	Filtration and separation (leachate)
11	Below waste	GN	Drainage (leachate)
12	Below waste	GM	Barrier (leachate)
13	Below waste	GTR	Separation (protection)
14	Below waste	GG	Reinforcement
15	Below waste	GP	Leachate removal

Geomembranes (GM)

The geomembranes used for liners beneath the waste have four very different requirements compared to the geomembranes used in covers above the waste. These greatly influence the overall geomembrane material selection and subsequent design. They are as follows:

- Geomembranes in the liner system are in contact with leachate, whereas those in the covers are only in contact with percolating water from rain and snow melt.
- Geomembranes in the liner system are under considerably higher compressive stresses compared to those in covers.
- Geomembranes in the liner system should be subjected to large deformations of the subgrade, whereas geomembranes in covers generally undergo large deformation due to the underlying waste.
- Geomembranes in the liner system can only be removed at enormous expense of removing the waste whereas those in the covers are more accessible for maintenance and possible remediation.

As suggested above, the geomembranes of the liner system beneath the waste must be chemically resistant to the leachate generated at the site. If aggressive leachates are present or an unknown synergy of chemical species is possible, proper tests will be necessary to select the proper polymer. Subsequent to the selection of polymer materials, a number of index strength tests (on both sheet and seamed material) and several performance strength tests are necessary. In this latter category, wide width tensile test, friction test and anchorage tests are usually necessary. To determine the effectiveness of the geomembrane as a hydraulic barrier, water vapour transmission tests are sometimes necessary. More significant, however, are the solvent vapour transmission tests depending on the nature of the leachates. Numerous other tests such as soil burial, stress crack resistance for semi-crystalline polymers, plasticizer leaching, dimensional change, ply adhesion of reinforced geomembranes, etc., are required on a site-specific basis and on a material-specific basis. Geomembranes are usually made from HDPE (most frequently) and PVC materials.

For the geomembrane located above the waste, chemical resistance is usually not a major factor. Strength tests will include index tests and wide-width tests, in addition to axisymmetric strength. The latter test should be performed on both sheet material and seamed material. Geomembrane seam tests, shear and peel tests are also required. It is also necessary to evaluate the vapour transmission of the geomembrane which would generally be methane vapour transmission above the waste. Geomembranes for this function are usually made from PVC (most frequently), LDPE and PP.

Geotextiles (GT)

Geotextiles in waste containment serve as leachate and water filters, drains for landfill gases and provide protection as a separator. Depending on their location in the facility, numerous tests are required. For geotextiles placed beneath the waste, chemical resistance tests via a simulated immersion procedure must be performed. Various index strength tests as well as wide-width strength and friction tests between the various interfaces involved are necessary. When the geotextiles serves as a filter, it is essential to evaluate parameters like permittivity, opening size and clogging (both particulate and biological). For geotextiles positioned above the waste, simulated testing just as mentioned above is necessary, with the exception of the chemical resistance test; in the case of geotextiles used as gas drainage layer, methane transmissivity test is additional. Geotextiles are usually made from PP, PVC and PE materials.

Geonets (GN)

A granular soil (sand or gravel) can be used for leak detection double liner systems. However, the current practice is generally to use a geonet as shown in Fig. 11.13. The reasons are as follows:

- It saves air space, as no perforated pipe system is necessary and the danger of damage to underlying geosynthetics is avoided.
- Stability on side slopes can be achieved and construction is rapid and straightforward.

Geonets, which are used as leak detection or primary leachate collection layer, must obviously have adequate chemical resistance against the leachate they transmit. Strengthwise, the compressive behaviour against rib collapse must be evaluated. Since long time durations are involved, compression creep is also a consideration. Most important is the evaluation of the transmissivity or in-plane flow behaviour under site-specific compressive stresses and hydraulic gradient. Its evaluation should use the simulated cross-section of materials above and below the geonet. In this manner the intrusion of the overlying and underlying materials can be properly evaluated. If geonet joining is required, the seam strength must be assured. Furthermore, the manner of seaming must not restrict the in-plane flow capabilities of the geonet. Friction testing of all interfaces must be performed. Geonets are usually made from HDPE or PP.

Geopipes (GP)

Perforated plastics pipes made from PVC or HDPE are commonly used as leachate collection systems. The interconnected system leads directly into down-gradient sump for removal and proper treatment of the leachate. Chemical resistance as well as adequate compressive strength must be assured. As with other components beneath the waste, creep deformation must be kept within tolerable limits. Since the pipes are usually laid in a network pattern, numerous fittings and joints are required. These are important details and must be evaluated with the utmost care.

Geocomposites (GC)

Geocomposite drains used as surface water collectors above the waste require the same tests as the geonets below the waste, but generally do not require chemical resistance evaluation. Additionally, the compressive strength requirements of these geocomposite drains are not as severe as those for geonets placed beneath the waste. Frictional characteristics must be evaluated, particularly when the closure side slopes become long and steep.

Geogrids (GG)

When used within the enclosed boundaries of the waste facility, as with vertical or horizontal extensions of an existing landfill, chemical resistance of geogrids must be necessarily assured. Strength tests should include junction and wide-width tests of short-term nature, as well as creep strength tests. It is also necessary to perform friction and anchorage tests. If connections are involved, the proposed joining method should be evaluated while in tension for both short- and long-term (creep) behaviour.

Geogrids outside of the waste enclosure (for example, for stabilization of ramps, berms and side slopes), require the same tests as just mentioned with the exception of chemical resistance. Also, geogrid joint tests may be avoided since the geogrids are usually configured to avoid connections. Geogrid creep behaviour is generally not critical, since stability problems are usually minimized after the waste is deposited and passive pressure is mobilized.

Geosynthetic clay liners (GCL)

They consist of factory-manufactured dry bentonite clay liners sandwiched between geotextiles or attached to a geomembrane. They can be constructed with lightweight equipment, minimising the risk of damage to underlying components, and can easily be placed on side slopes.

Considering both the specifications/guidelines presented earlier and the liner and cover systems shown in Fig. 11.13, geosynthetic materials can be considered as better alternatives to natural materials. Figure 11.14 shows the difference in thickness of the components that leads to savings in space in the case of liners if geosynthetics are used. It may be noted that the savings in space is considerable and the additional volume made available can accommodate more waste before closure. The following are the replacements that are considered appropriate:

- Geomembrane and geosynthetic clay liner in place of clay soil
- Geonet in place of drainage gravel
- Geocomposite in place of drainage gravel
- Geotextile in place of filter sand
- Geopipe in place of concrete pipe

11.5 Barrier Walls for Existing Landfills and Abandoned Dumps

The cut-off and/or containment of laterally-flowing liquids from landfills and impoundment reservoirs generally utilize some type of vertical wall. The most common walls are constructed using a slurry-supported trenches, subsequently backfilled with soil–bentonite, soil–cement, cement–bentonite or soil cement–bentonite. The installation, inspection and durability of such walls are a cause for concern. A different or a complementary strategy uses a geomembrane alone, or in combination with any one of the standard backfill materials, to provide the degree of completeness and environmental safety/security that most of these sites warrant.

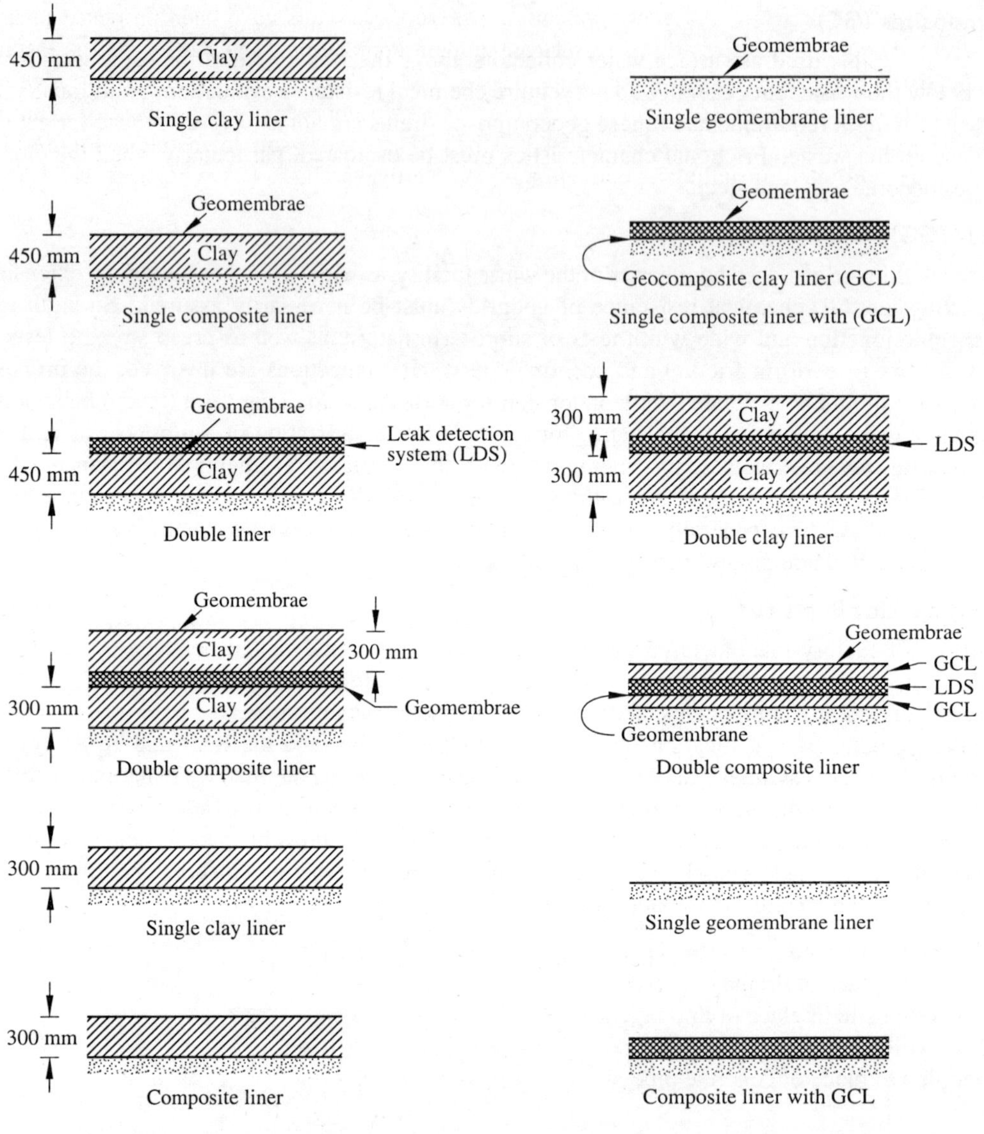

Fig. 11.14 *Landfills (typical cross sections with and without geosynthetic material)*

Barrier walls containing a geomembrane can be useful in any application where the flow of water or other liquids is to be controlled. The use of a geomembrane in the wall ensures the extension of an extremely low permeability material. In the light of the numerous concerns found in conventional solutions, it is clear that barrier walls that include geomembranes provide an added factor of safety for almost any construction application. The walls are generally installed by placing individual panels that are connected together. There are several different interlocking connections available and there are different ways to seal a connection. The geomembranes used for the panels have almost always been

HDPE. There are a variety of different applications where these walls have been proposed or used. Figure 11.15 shows a typical hazardous waste containment application. The barrier wall is installed at a depth that is either well below the plume of waste or keyed into the aquitard. This arrangement (vertical barrier and capping) is frequently used to create a watertight barrier around an existing unlined landfill when the flow of the water table can disperse the leachate in the soil downstream. Figure 11.16 shows a typical method of installing a geomembrane as a vertical wall.

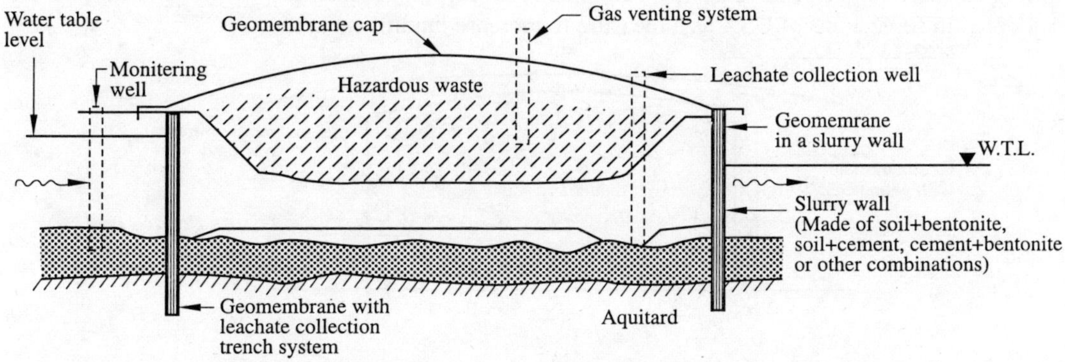

Fig. 11.15 *Confinement of contaminated soil with geomembrane vertical wall with closure cap*

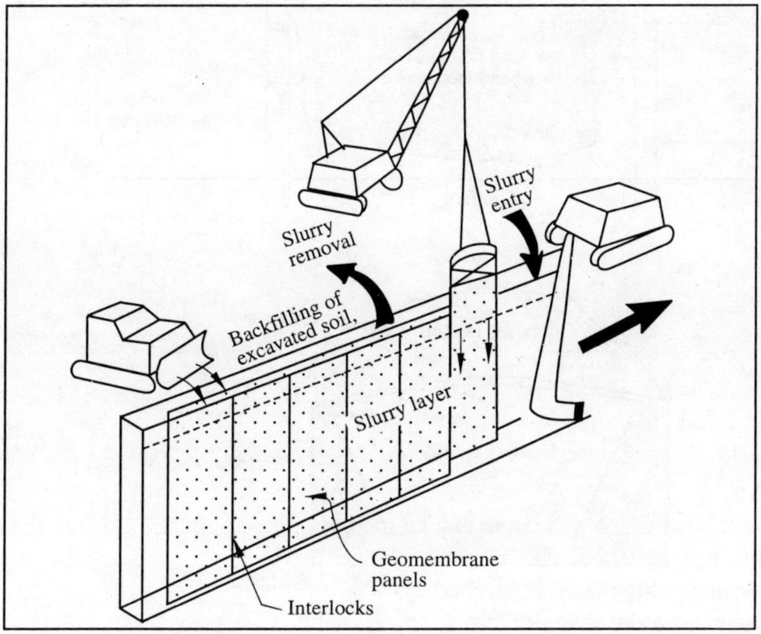

Fig. 11.16 *Installation of geomembranes*

Due to the unknown extent and nature of abandoned waste dumps, one strategy is to leave the waste material in place and contain it with a vertical barrier. Koerner (2000) highlighted some possible uses of geosynthetics in this regard. Figure 11.17 shows the potential applications. Figures 11.17(*a* & *b*) show the uses of geomembrane, geosynthetic clay liner (GCL) and geonet as barriers/facilitators in containing impact of abandoned dump. Figure 11.17*c* shows how an existing landfill is covered with geomembrane, and how geosynthetic reinforcement is provided in such a way that additional landfilling capacity can be achieved. This technique is called piggyback expansion of landfills and has been adopted in some areas of USA as a measure to increase landfill capacity.

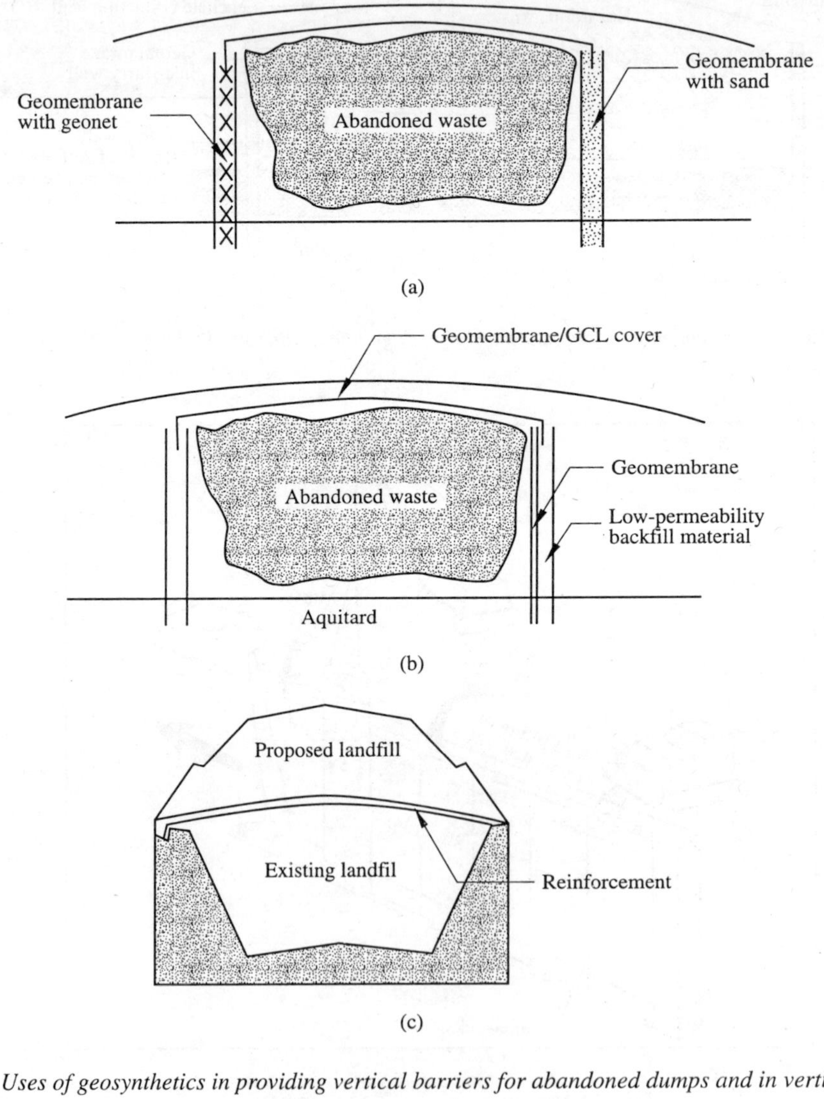

Fig. 11.17 *Uses of geosynthetics in providing vertical barriers for abandoned dumps and in vertical expansion of landfills (Koerner, 2000)*

11.6 Some Issues/Concerns in Usage

The containment of solid (and liquid) waste facilities is greatly dependent on the proper and long-term functioning of geosynthetic materials made from polymers. Since geosynthetic materials in use have only a relatively short history (of past experiences), many concerns are voiced which should certainly be addressed. Some selected and more common concerns regarding the use of geosynthetics for waste containment applications are presented below.

1. Chemical resistance

Whenever a polymeric material interfaces with a leachate or other aggressive liquids, the chemical resistance of the resin or compound must be examined. The heterogeneity of the leachate at any given time and the possible variations of it over the lifetime of the facility, make the development of standards and adherence to guidelines for uniform incubation and testing procedure for all types of geosynthetics a difficult task. Standards are available from ASTM, EPA or European Norms, which should be followed.

2. Geosynthetic seams

The performance of a facility depends on the effectiveness of the seaming in the field. All seaming methods are based on heat sources softening or melting the geosynthetics to be welded together. Many destructive and nondestructive geomembrane seam testing methods are available in literature in the form of codes, standards and guidance documents from organizations as well as manufacturers.

3. Clogging of filters and drains

Among the various possible strategies that could be considered in the light of geotextile filter clogging by particulates and bacteria in landfill leachates, the most prudent approach seems to be a design which allows for some degree of access and continuous maintenance via backflushing through the system.

4. Long-term degradation

Geosynthetics are quite difficult to place, but their durability in a buried environment is far longer than ever suspected. When properly placed and seamed, the degradation of any geosynthetic material must come from molecular chain breakdown of the polymeric resin itself or bond breaking and reactions with the compounding material. Accelerated ageing tests at elevated temperatures are to be undertaken. Even exhumation of geosynthetics from old facilities has been conducted to understand the problem better. As of now there are many geomembranes in excess of 30 years of successful service. Sophisticated laboratory tests have shown that a service life in excess of 100 years is possible for the most common resins and geosynthetic materials made them.

11.7 Conclusions

Use of geosynthetic materials for waste containment has vast potential. The quality of the products and of their installation must be verified at all times through strict quality assurance and quality control plans. Only products from reputable, experienced manufacturers should be used. Installation plays a key role in the performance of a facility. Hence discussion among the contractors, manufacturers, engineers and experts is necessary to optimise the use of geosynthetics.

Bibliography

[1] Central Pollution Control Board (2000) *Criteria for Hazardous Waste Landfills*, Ministry of Environment and Forests, Government of India.

[2] Daniel, D.E. (1987) Earthen liners for land disposal facilities, *Proceedings Geotechnical Practice for Waste Disposal*, University of Michigan, ASCE, 21–39.

[3] Daniel, D.E. and Koerner, R.M. (1997) Construction quality assurance/quality control for landfills, in *Municipal Solid Wastes: Problems and Solutions*, (Landreth, R.E. and Rebers, P.A. Eds), CRC Press, Boca Raton, FL.

[4] Daniel, D.E. and Koerner, R.M. (1993a) Cover Systems, in *Geotechnical Practice for Waste Disposal*, Daniel, D.E., Ed., Chapman and Hall, New York, pp. 455–496.

[5] Daniel, D.E. and Koerner, R.M. (1993b) *Quality Assurance and Quality Control for Waste Containment Facilities*, U.S. Environmental Protection Agency, EPA/600/R–93/182.

[6] Gartung, E. (1996) Landfill liners and covers, *Proceedings of First European Geosynthetics Conference*, Maastricht, Netherlands, Balkema, 55–70.

[7] Koerner, R.M. (1998) *Designing with Geosynthetics*, 4th Edition, Prentice-Hall Publ. Co., Englewood Cliffs, NJ.

[8] Koerner, R.M. (2000) Emerging and future developments of selected geosynthetic applications, Terzaghi lecture, *Journal of Geotechnical and Geoenvironmental Engineering*, ASCE, April, 2000, 293-306.

[9] Landreth, R.E. (1997) *Cover Systems for Waste Management Facilities in Municipal Solid Wastes: Problems and Solutions*, Landreth, R.E. and Rebers, P.A., Eds., CRC Press, Boca Raton, FL

[10] MoEF (2000) Municipal Solid Wastes (Management and Handling) Rules, 2000

[11] Sharma, H.D. and Lewis S.P. (1994) *Waste Containment Systems, Waste Stabilization and Landfills, Design and Evaluation*, John Wiley and Sons Inc. New York.

[12] Scuero, A.M. (1997) The use of Geosynthetic Materials for Waste Containment, *Geosynthetics Asia '97*, pp. A85–A92.

12 Natural Geotextiles

12.1 Introduction

The advent of geosynthetics in the recent past has brought in a new dimension to the solutions of various civil engineering problems. Economical, ecological and environmental considerations dictate that if natural materials that are biodegradable are in abundant supply, their use is justified in engineering solutions. This is especially true in countries such as India and Bangladesh where some of these materials are readily available. One can envisage a wider application of natural geotextiles such as jute, coir, bamboo, etc. as replacements to geosynthetics to solve some civil engineering problems. In some applications, both natural and synthetic materials can be used together to ascertain that some mechanical properties are sustained long after degradation of natural materials have occurred. This chapter presents properties of some natural geotextiles such as jute, coir, bamboo, etc., and their relative advantages, disadvantages and application areas.

The introduction of synthetics in engineering has opened many possibilities in the form of new techniques, design and construction methods, especially in civil engineering. This development is more or less based on the knowledge of the behaviour of synthetic materials. Natural fibres can very well be used in applications like erosion control, separation, filtration and reinforcement when the required life-span is short. Properties such as the ability of mature fibres to absorb water and their biodegradability with time give them an edge over synthetic geotextiles for erosion control purposes. The drapability factor of natural geotextiles (due to their flexibility) allows them to conform closely to the terrain. Bamboo dowels, due to their high tensile stress, can be used efficiently to prevent landslides/slope-failures. The main challenge while using natural fibres is to protect them from biodegradation and attack from micro-organisms. This can be done to some extent by coating with polymers, bitumen, etc.

12.2 Natural Fibres as Geotextiles

Natural fibres can be grouped into three classes.

- *Blast fibres:* These are extracted from stem of plants. Jute, flax, hemp and ramie come under this category. Flax, hemp and ramie are used in twines, canvases, fishnets, etc. Jute is used in sacking, carpeting, sand bags, concrete curing, etc. It is also used for mulching and weed control.
- *Leaf fibres:* Sisal comes under this category. Leaf fibres are used as cordage or rope making material, for making mats, etc.
- *Seed/fruit fibres:* Cotton and coir comes under this category. Coir has similar end uses as leaf fibres.

Natural fibres can be considered suitable as geotextiles if (i) they have reasonably good mechanical properties and (ii) are reasonably resistant to microbial attack. Based on the above considerations, coir and jute fibres are adjudged suitable for use as geotextiles. Besides these materials, bamboo and timber can be used as natural reinforcement in many applications due to their high tensile strength (in the range of 800 kN/m^2) and compressive strength (approximately 300 kN/m^2). Straw can be used as random reinforcing material. Considerable information on the use of jute geotextiles is available from agencies such as Indian Jute Mills Association (IJMA), Indian Jute Industries' Research Association (IJIRA), Kolkata. With regard to the use of coir as civil engineering material, Rao and Balan (2000) prepared an excellent compilation that may be referred to for more details. A number of case studies of successful applications of these materials are available with the above agencies and are cited in the above references. In the following sections a brief review of these materials, their properties and applications is provided.

12.3 Factors Governing the Usage of Natural Geotextiles

Reinforcement

The tensile stress–strain behaviour of geotextiles and the soil–geotextiles interface frication are of importance in estimating the performance of geotextiles as reinforcement in civil engineering applications.

Separator in pavements

During the installation process, the fibers are subjected to dynamic impact of moving vehicles against the crushed rock boulders below. Therefore properties such as puncture resistance, tensile strength and resistance to sudden impact of the geotextile need to be optimised for improvement in the pavement performance.

Erosion control

Some requirements for erosion control are:

Coverage

The surface should be covered as much as practicable, with scope for sufficient passage of light for growth of vegetation.

Drapability

This property will help the material develop intimate contact with soil so that run-off does pass under it, thus arresting movement of soil, seed and nutrients.

Water absorbency

This will help the material to hold water, thus reducing the run-off velocity.

Tensile strength

The material should have sufficient strength to withstand handling, transportation and installation operations.

In addition to this, the site of application must be satisfactory from geotechnical considerations, otherwise minor/major geotechnical/structural corrections may have to be under taken later.

Filtration

In filtration applications, the geotextile is placed as an interface between soft clays of very low permeability, and free draining media like sand and acts as a separator and filter. The permeability of geotextiles must be significantly greater than that of clayey soil and nearly equal to that of free draining medium. To act as good horizontal and vertical drains, the geotextiles must be thick enough and/or have suitable in-plane permeability or transmissibility characteristics.

Agricultural applications

The fibre should have sufficient pores for aeration and strength to withstand handling stresses; be non-toxic to the plants and good enough to conserve moisture and heat for germination; and protect the seedlings from the scorching sun and lashings of rain during the early stages.

12.4 Jute Fibres

Jute is one of the most valuable natural fibres produced extensively in India and Bangladesh. Under the product name *soil saver*, it is been an export commodity for many years. It is often called *golden fibre* in Bangladesh. The chemical composition of Jute is as follows:

- Alpha cellulose (61%),
- Hemi cellulose (24%, including uronic acid derivatives),
- Lignin (11.5%)
- Fatty acids and waxy substances with some nitrogenous and mineral matter (3.5%)

12.4.1 Various products of jute

Jute-based geotextiles are evolved using a process of mixing a blend of yarns. Jute products are woven or non-woven, with a very wide range of engineering properties to suit different requirements. The roll-width of jute fabric is normally up to 3 m long and length is up to 100 m. Geojute fabrics, jute meshes, jute mats, jute nets, jute drains, etc., are various types of jute products. Typical properties of fibres and fabrics are presented in Tables 12.1 and 12.2.

Aziz and Ramaswamy (1991) brought out the advantages of jute fabrics from a study of compaction characteristics, load–settlement relationships, unconfined strength tests, CBR tests, plate load tests and response of subgrade/sub-base profile under static and dynamic loading. Figures 12.1 and 12.2 show some of the results obtained. Figure 12.1 shows the response of soil reinforced with jute fabric. The dynamic load applied was 8 kN, the simulated contact pressure was 255 kN/m^2 and up to 1000 load applications were made. Clay with water content of 40% was used as subgrade soil. The response was measured in terms of rut depth and load repetitions. It was noted that there is considerable reduction in rut depth with the use of jute geotextiles as well as with increase in aggregate thickness. The results also indicated that for the same rut depth, the number of load repetitions increased. Results of static loads presented in Fig. 12.2 indicate that when jute fabric is used as subsurface material between the subgrade and the aggregate material, the load-carrying capacity increased at the same settlement level.

Table 12.1 *Typical properties of jute fibres (Aziz and Ramaswamy, 1991)*

Properties	Range of values
Fibre length (mm)	180–800
Fibre diameter (mm)	0.10–0.20
Specific gravity	1.02–1.04
Bulk density (kg/m^3)	120–140
Ultimate tensile strength (N/mm^2)	250–350
Modulus of elasticity (kN/mm^2)	26–32
Elongation at break %	2–3
Water absorption %	25–40

Table 12.2 *Typical properties of jute fabrics*

Properties	Range of values
Thread diameter (mm)	1.75–1.85
Mesh size (cm^2)	3 × 3 (double)
Weight (g/m^2)	680–750
Elongation at break (wet) %	15–20
Trapezoidal tear strength (N)	300–350
Permeability (cm/sec.)	
- under unstressed conditions	$> 10^{-2}$
- under all round pressure of 500 kN/m^2	$> 10^{-3}$–10^{-4}
Grab tensile strength (wet) N	800–900
Tensile strength (kN/m)	15–25
Puncture resistance (N)	350
CBR pushing through resistance (N)	1300–3000
Cone drop penetration (mm)	8–10

Figure 12.3 shows the observed behaviour schematically. The above results indicate that common road defects such as rutting that arise during the early stages in the case of roads on weak subgrades can be effectively prevented if jute geotextiles are used in the initial stages.

Srinivasa Murthy et al. (1994) reported the advantages of jute fabrics in filtration and drainage applications from the results of long-term tests conducted which involved monitoring at different hydraulic gradients. They suggest that even after degradation of the jute geotextiles, a filter cake was formed at the location where jute geotextile was present earlier and this ensured a comparatively better performance of the filter to the conventional granular filter which is likely to get choked over a short term. Rao (1996) presents a number of case studies in which jute geotextiles were used in many applications such as pavement separator, filter (behind retaining walls), and for erosion control. Table 12.3 presents a comparison of properties of jute fabrics with other commonly used geotextiles. It shows that the jute is heavier, thicker and its elongation at break lesser than synthetics. However, the other properties are comparable.

12.4.2 Advantages and disadvantages

- It is economical and renewable resource.
- It is available abundantly especially in India.

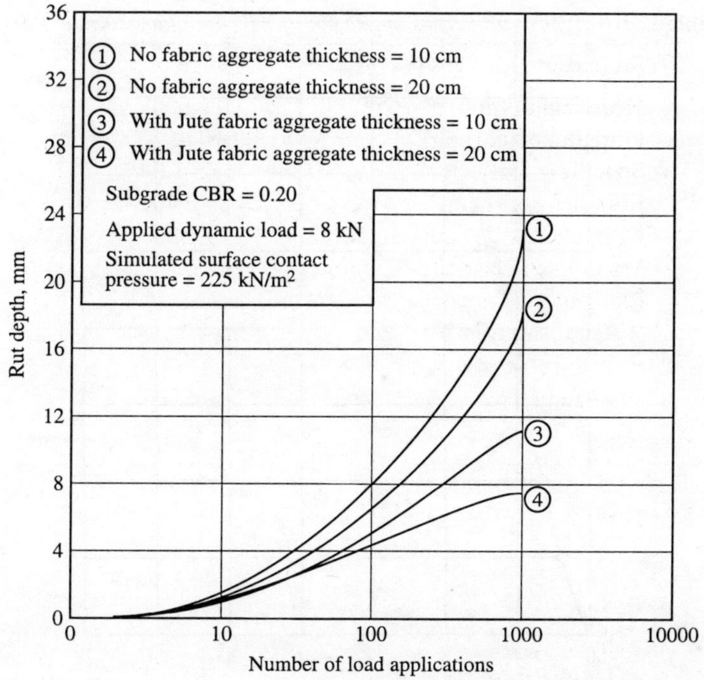

Fig. 12.1 *Typical rut depth versus number of load applications in dynamic load testing using jute fabrics*

- It is an environment-friendly natural product.
- It is biodegradable and acts as a nutrient for vegetation.
- The drapability of jute is very high and increases by 25% when wet.
- It has good fibre structure, and when used in combination with soils improves the hydraulic properties such as drainage and filtration.
- It improves the mechanical properties of soils.
- As jute fibre can absorb water five-and-a-half times of its weight, geojute helps to conserve moisture (up to 40–50%) in the soil on which it is laid, which helps the growth of vegetation.
- Jute fabric has sufficient strength and non-toxic properties desirable for use in agricultural applications, whereas synthetic material in such applications would have an adverse effect. This aspect has led to its greater use in erosion control and such similar measures.
- Jute is a bad conductor of heat and electricity and serves as insulation material and protects the soil surface against both heat and cold.
- When treated with appropriate preservatives, it is also effective over longer terms though this aspect needs more research and analysis through case studies.
- It degrades by microbial attack quickly, especially in humid regions.
- The relationship between the rate of decrease of strength under different soil and environmental conditions needs to be established.

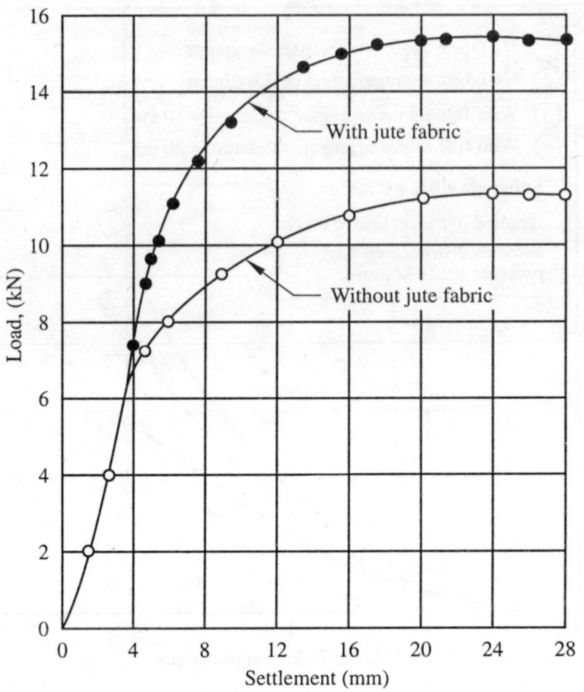

Fig. 12.2 *Load–settlement relationship with and without jute*

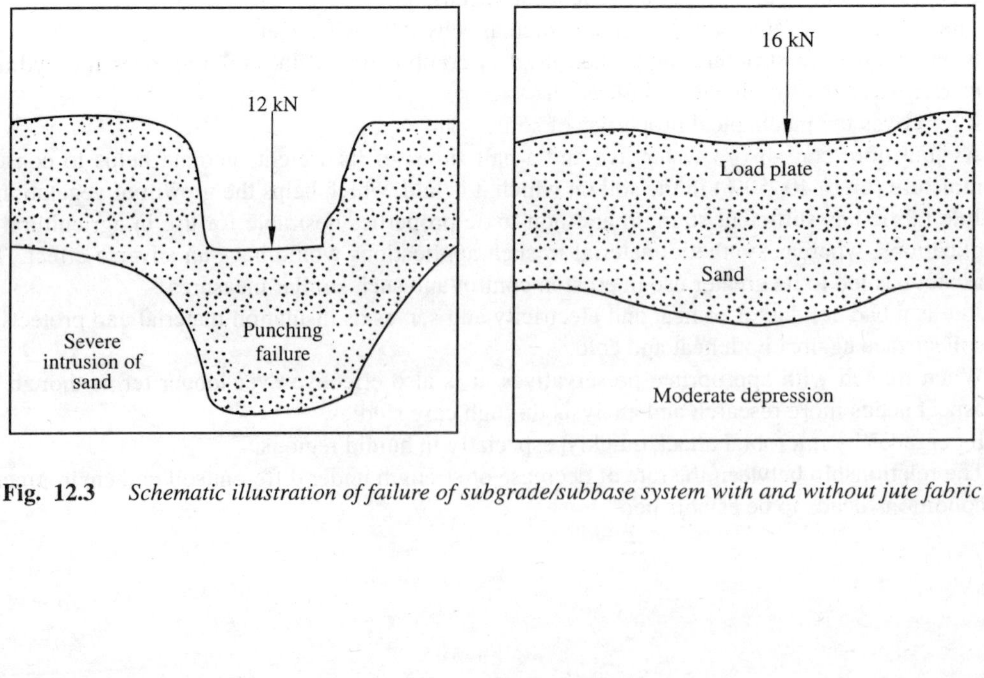

Fig. 12.3 *Schematic illustration of failure of subgrade/subbase system with and without jute fabric*

Table 12.3 *Comparison of properties of jute fabrics with other commonly used geotextiles (Aziz and Ramaswamy, 1991)*

Trade names	Weight (gm/m^3) ASTM D-1910 -64	Thickness (mm) ASTM D-1777	Grab tensile strength (N) ASTM D-1682	Elongation at break ASTM D-1682	Trapezoidal tear strength (N) ASTM D-2263	Peremeabi-lity (cm/s) falling head	Type
Propex 4545	153	—	400 (min)	50 (min)	—	2×10^{-2} (min) 3×10^{-2}	Nonwoven
Mirafi 600X	—	—	1335	—	534	0.01	Woven
Typar	136	15 mils	580	62	312	0.02	Nonwoven
Supac 5-P	180	15 mils	556–670	80	325	0.05	Nonwoven (polypropy-lene)
Petromat	146	—	512	65	—	—	Nonwoven
Terram 140	280	1.1	1128	150	34	0.072	Nonwoven 75% poly-propylene 25% Nylon
Jute fabric	680–750	1.75–1.85	800–900	15–20	300–350	0.02–0.04	Woven

12.4.3 Enhancing the durability of jute

- The average life of jute when used in soil (approximately 2 years) may be enhanced several-fold by proper treatment with bitumen coating.
- Woven jute fabrics can be treated chemically for retarding rotting.
- The biodegradation of jute can be effectively retarded with the surface application of phenol formaldehyde resin or a coating of polypropylene, polyethylene, etc.
- Blending the jute fibres with synthetic fibres such as polypropylene, polyethylene, etc., is also used.

12.4.4 Applications

For reinforcement

- Over soft soil for unpaved roads, temporary approach roads
- Over unstable slopes as vegetation cover
- To stabilize slope of temporary bunds, ash ponds, etc.

As separator

- Between subgrade and ballast in railroads
- Between drainage layers in poorly graded filter blankets

As filter

- In place of granular soil filters
- Beneath stone-base in unpaved and paved roads and air fields
- Beneath ballast under railroads
- Around stone aggregate in drain
- Around perforated drain pipes
- Beneath rip-rap

In drainage

- To dissipate seepage water from exposed soil or rock surfaces
- Drainage of pavement layers using geotextiles

Others

- In erosion control
- In soil conservation
- Concrete curing
- Silt fences

The Bureau of Standards of India formulated specifications for woven jute geotextiles; it is given in IS 14715(2000) for different applications such as soil slope protection along hillside, roadside and railway tracks, river embankment protection and other applications involving filtration and separation functions.

12.5 Coir Geotextiles

Coir is a natural product, which is available in India at low cost. Coir is obtained from husk of coconut. Many products of geotechnical interest are manufactured from coir fibres. An excellent treatise on coir geotextiles is compiled by Rao and Balan (2000) for Kerala State Coir Corporation Limited. It contains methods of making coir geotextiles, properties, applications and a few case studies. For more details readers may refer to this compilation. Chemical composition of coir fibre is as follows:

Table 12.4 *Chemical composition of coir fibre*

Lignin	41–45%
Cellulose	36–43%
Hemi-cellulose	0.15–0.25%
Pectin	4%
Water solubles	5.25%
Ash	2.22 %

12.5.1 Products of coir

India produces the best quality of coir fibre from the coconut plant. Two kinds of coir fibres are available in India, viz. white fibre and brown fibre. Both are derived from extraction process. White coir is made from mature green coconuts by subjecting the husk to retting (a biological process which takes about 8 to 10 months in which coconuts are soaked in water to remove certain natural chemicals so that defibering is possible), followed by beating of husks with mallet against a log by a manual labour for thrashing out the coir pith. Brown coir is produced from dry/semi dry coconuts by a mechanical process after soaking the husks for brief periods in water. Both types can be used for making geotextiles, woven or non-woven, depending on the end use. Wide varieties of geotextiles ranging in densities from 400 to 1400 g/m^2 can be developed from coir.

Table 12.5 *Typical properties of coir fibre*

Properties of single fibre	Range of values
Length (cm)	15–20
Density (g/cm^3)	1.40
Tenacity (g/tex)	10.0
Breaking elongation (%)	30
Moisture regained at 65% R.H. (%)	10.5
Swelling in water (diameter) %	5.0

Table 12.6 *Typical properties of coir mat*

Specific gravity	1.402
Mass per unit area	1.396 g/m^2
Thickness at 2 kPa	7.2 mm
Stiffness	300 mg-m
Tensile strength	37 kN/m
Secant modulus at 10% strain	110 kN/m
Tensile toughness	8.21 kN/m

12.5.2 Advantages and disadvantages

Advantages

- The reinforcement products made of coir are comparatively stronger and stabler than jute and other materials.
- The initial strength, stiffness and hydraulic properties of coir reinforcement are almost comparable to those of similar products made from polymer materials.
- When the requirement of reinforcement is for short duration, coir products can be a good choice.
- By chemical treatment and polymer coating, the life of coir products can be improved.
- Wide variety of geotextiles ranging in densities from 400 to 1400 gm/m^2 can be developed.
- It can be laid on any surface owing to its flexibility and is hence useful for geotechnical purposes.
- Use of coir geotextiles is economically viable and is environment-friendly for protecting slopes in erosion prone areas.
- It takes 15 times longer than cotton and 7 times longer than jute to degrade.
- Very low raw material price
- No chemicals, lubricants or emulsions are added at the time of spinning. Hence there are no toxic elements on the yarns after it is spun and is safe for use.

Disadvantages

- The coir products are vulnerable to different degradation actions in the field such as those induced by installation damage, temperature fluctuations, chemical actions, hydrolysis, ultraviolet degradation and other degradation processes.
- The reinforcement products of natural coir fibres develop their ultimate strength at higher strains as the modulus is low. Hence their capacity is not fully utilized if there is a restriction on the maximum deformations that a structure can undergo. This can be overcome by coating them with thermosetting polyester to increase their strength and stiffness.
- The areas of application have been limited till date, and are mainly in the field of soil erosion control.
- Coir fibres, extracted from dry coconut husk exhibit a wide range of dimensions which makes conventional spinning process difficult.

12.5.3 Preservation

A coating of thermosetting polyester improves the durability, strength, stiffness of coir fibres and decreases its water absorption. The effectiveness of polymer coating was found to depend on the pretreatment methods used before applying the polymer coating. Samples with copper pre-treatment have given better results.

Considerable work has been conducted on the use of coir in India and a number of applications have been reported by Rao and Balan (1997, 2000), Beena (1997), Rajagopal and Ramakrishna (1997) and others. Figure 12.4 shows a typical result from Rao and Balan (1997). It can be noted that coir-mixed soil showed higher initial tangent modulus compared to unreinforced sand.

12.6 Bamboo/Timber

Strength and deformation characteristics of several natural materials such as bamboo, timber and fibres compare very favourably with synthetics. It fact, with regard to the deformation modulus and creep, natural materials such as bamboo/timber are superior to synthetics. Datye (1988) presents a very good account of the uses of bamboo and timber in Indian conditions along with their cost advantages. Typical properties of these materials are given in Table 12.7.

Table 12.7 *Typical properties (Datye, 1988)*

Timber (40 mm diameter)		Bamboo dowels	
Weight (gm/m^2)	1430	Allowable tensile strength	80,000 kPa
Strength at 2% strain (kN/m)	25	Allowable compressive strength	30,000 kPa
Modulus at 2 % strain (kN/m)	1250		

Matsunaga et al., (1988) describes a case study of reclamation of soft ground in which sheets of polymer nets were used for segregating the existing ground and the embankment material with bamboo cross beams to support the vertical load from the embankment. Tension tests on bamboo indicated a

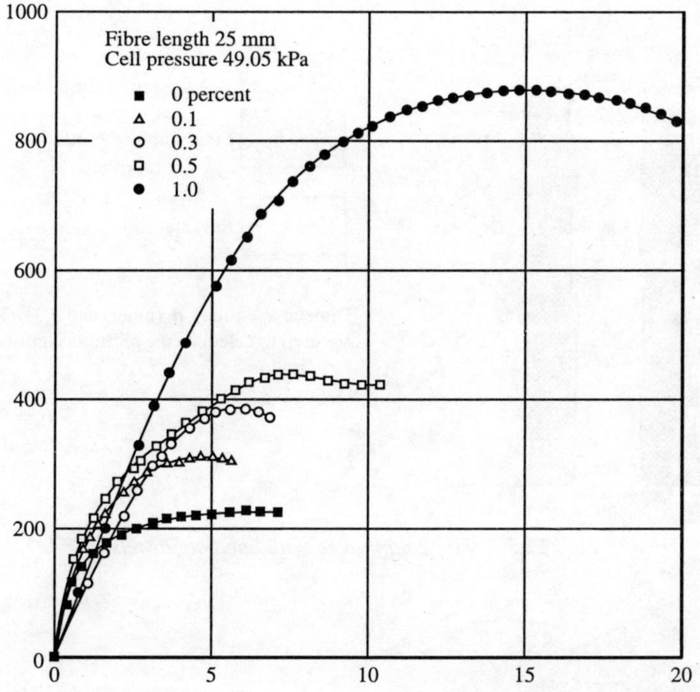

Fig. 12.4 *Effect of coir fibre on deviatoric stress from the study of Rao and Balan (1997)*

mean tensile strength of 160000 kPa and a modulus of elasticity of 87.8×10^5 kPa, and it was found that the age of bamboo has little effect on the mechanical properties. Poorooshasb et al. (1988) presented a case study of stabilization of a residual soil slope in Brazil using bamboo. The authors presented a design methodology based on which the spacing and length of bamboo dowels can be arrived at to restrict the slope movement (see Fig. 12.5). The authors also suggest that the dowels could be encased in a protective sleeve in which lime/cement stabilized soil could be used as shown in Fig. 12.6. It is reasoned that this arrangement prevents the dowel from acting as unsupported beam and also prevents local buckling.

12.6.1 Advantages and disadvantages

- Bamboo dowels can be efficiently used as reinforcement in preventing downhill movement of slopes.
- Cost advantages are significant.
- Material characteristics are superior in the respect of deformation modulus.
- The small timber cribs are cost effective in comparison to concrete cribs.
- By judicious design, a combination of small timber ties and crib facings could replace gabions.
- The small-sized timber reinforcements are suitable for anchored constructions.
- The extensions are smaller and the form of joints used impart adequate flexibility.
- Timber reinforcements compare very effectively with some synthetic grids.
- They can replace steel rods in short pile reinforcements of bases of embankments on soft soils.

The disadvantage is that they are prone to attack from microorganisms.

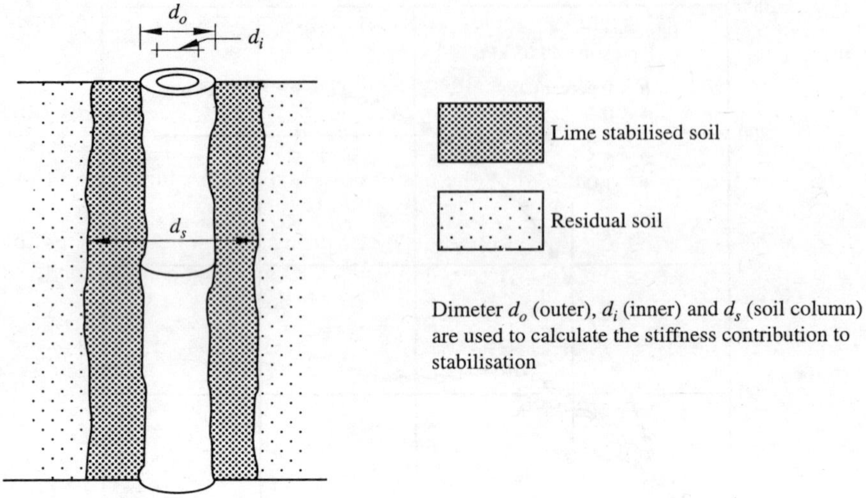

Fig. 12.5 *Slope reinforced with bamboo dowels*

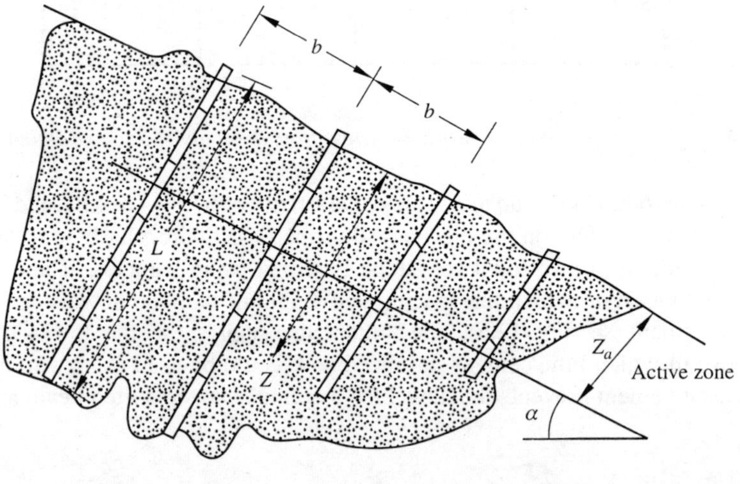

Fig. 12.6 *Encasement of dowels in a protective sleeve in which lime/cement stabilized soil is used*

12.6.2 Preservation methods

The following are some of the preservation methods suggested in literature.

- Impregnation with water and oil-borne preservatives
- Impregnation with synthetic polymers
- Coating with cold setting liquid resins
- Coating with synthetic melts
- Encasement in concrete or cement composites (e.g. Ferro cement)

- Water soluble preservatives and fungicides (successfully used in the past) for treatment of timber and bamboo
- Treatment with water repellents such as silicone and PVC emulsions
- Using colloidal silica admixtures, very tough and durable cement mortar coats of 6 to 8 mm thickness can be produced.
- Corrosion resistance can be achieved by the use of galvanized iron, aluminum, or high modulus synthetic meshes.
- By using a suitable coating and providing joints in the protective annulus, it is possible to ensure that the coating would be relieved of stress and therefore cracking can be avoided.

12.7 Combination of Geotextiles

Attempts have been made to combine the relative advantages of coir, jute or synthetic materials by blending these materials in appropriate proportions. Jute fibres exhibit moderately high modulus as well as tenacity and very low elongation at break, whereas coir fibres behave exactly in the opposite manner namely moderately low modulus, low tenacity and very high elongation at break (Benerjee and Unni, 1997). So, logically, they form ideal partners for blended product. This fact can be seen from the results of Aziz and Ramaswamy (1991) which are presented in Figs. 12.7, 12.8 and 12.9. The results in terms of static loading tests, sustained loading tests as well as dynamic loading tests on jute– mat– coir grid clearly show that the combination of these materials is definitely superior to individual materials.

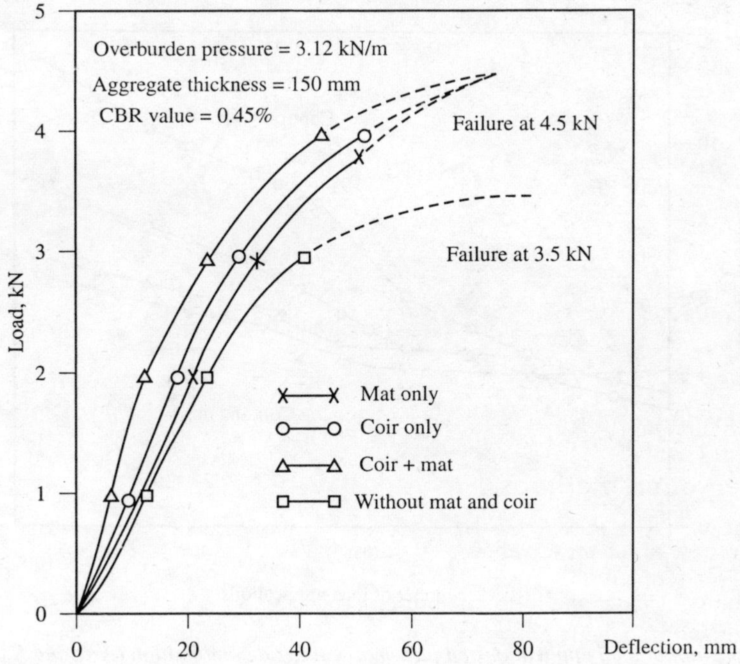

Fig. 12.7 *Load tests on jute–mat–coir grid combinations (Aziz and Ramaswamy, 1991)*

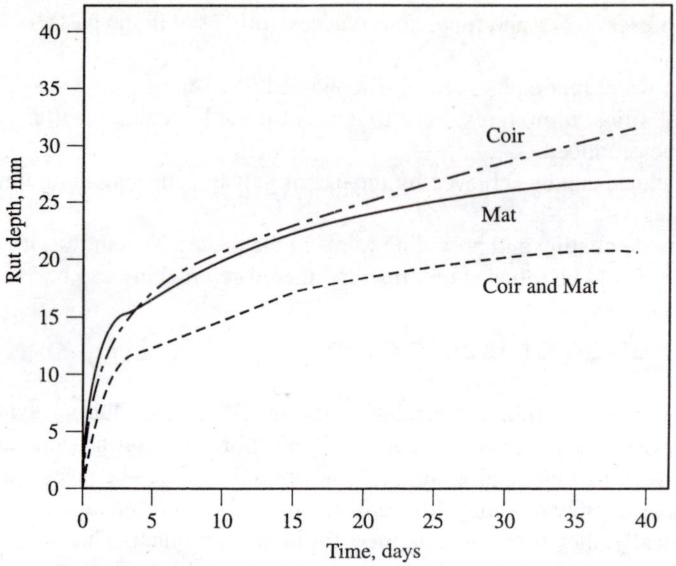

Fig. 12.8 *Rut-depth–time relationship for sustained loading tests (Aziz and Ramaswamy, 1991)*

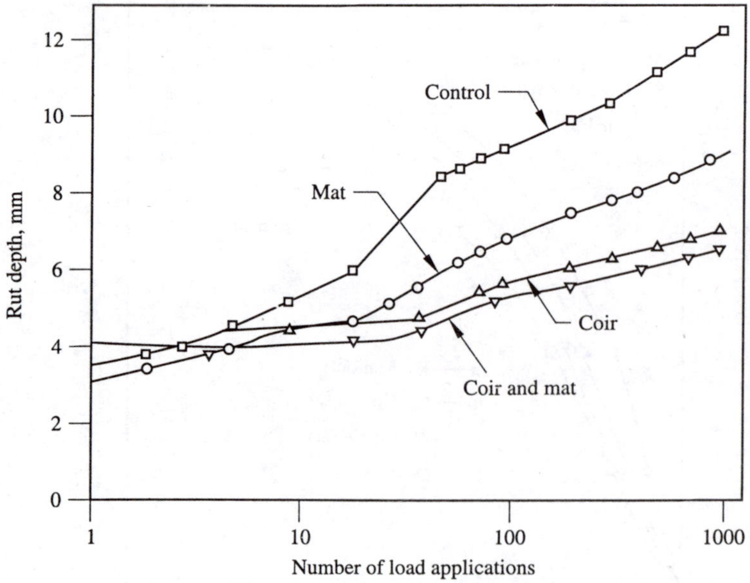

Fig. 12.9 *Rut-depth–no. of load applications on jute–mat–coir grid combination (Aziz and Ramaswamy, 1991)*

12.8 Concluding Remarks

In this chapter, some information on the use of natural materials such as jute, coir and bamboo in civil engineering is presented. These materials are economically, ecologically and environmentally suitable for some applications. They have good mechanical properties and can be made resistant to microbial attack with different treatments. A few specifications have been laid down by various agencies for augmenting their properties for specific purposes.

Bibliography

[1] Aziz, M.A. and Ramaswamy, S.D. (1991) Properties of jute geotextiles for civil engineering applications, *Textile Trends Journal*, Eastland Publications, Kolkata, 29–40.

[2] Banerjee, P.K. and Unni, E.B. (1997) Growth prospects of natural Geotextiles, *Geosynthetics Asia '97*, Bangalore, India.

[3] Beena, K.S. (1997) Use of bamboo and coir reinforcement—an approach, *Geosynthetics Asia '97*, Bangalore, India.

[4] Datye, K.R. (1988) Natural materials for soil reinforcement, *International Geotechnical Symposium on Theory and Practice of Earth Reinforcement*, Fukuoka, Japan, 69–74.

[5] IS:14715 (2000) Woven Jute Geotextile Specifications.

[6] Matsunaga, S., Watari, Y. and S. Isayama (1988) Reinforcement of very soft ground using bamboo cross beam with polymer net, *International Geotechnical Symposium on Theory and Practice of Earth Reinforcement*, Fukuoka, Japan, 425–430.

[7] Poorooshasb, H.B, Azevedo, R. and Ghavanmi, K. (1988) Analysis of slopes reinforced with bamboo dowels, *International Geotechnical Symposium on Theory and Practice of Earth Reinforcement*, Fukuoka, Japan, 467–472.

[8] Rajagopal, K. and S. Ramakrishna, (1997) Degradation behaviour of coir-geotextiles within clay soils, *Geosynthetics Asia '97*, Bangalore, India.

[9] Rao, G.V. and Balan, K. (1997) Reinforcing sand with coir fibre, *Geosynthetics Asia '97*, Bangalore, India, A183–190.

[10] Rao, G.V. and Balan, K. (2000) *Coir Geotextiles, Emerging Trends*, Publication of the Kerala State Coir Corporation Ltd., Alappuzha, Kerala.

[11] Rao. P.J. (1996) Proceedings, *National Seminar on Jute Based Geotextiles*, Central Road Research Institute, New Delhi, 1996.

[12] Srinivasa Murthy, B.R., Nagaraj, T.S. and Manoj, M. (1994) Jute geotextiles—an eco-friendly material, *Proc. of Indian Geotechnical Conf.*, Warangal, 281–290.

Index